The Chase for Zero Gravity

Einstein was wrong

AF625387

The most important book in the fight against the brainwashing within today's theoretical physics.

-It is easier to believe in a lie,
heard one thousand times,
than in a piece of truth,
heard for the first time.

Unknown

The figure is
dedicated to
the memory
of the photon.

This book is a translation of the Swedish book, ”Jakten på tyngdlösheten”, which should be consulted, when in doubt of the translation.

English edition: 9

Publisher: BoD · Books on Demand, Östermalmstorg 1, 114 42 Stockholm, bod@bod.se
Print: Libri Plureos GmbH, Friedensallee 273, 22763 Hamburg, Tyskland
ISBN: 978-91-7463-796-0

F I R S T =>

The book has now been around since 2011, and it has unfortunately turned out that no matter how easy it is to read and how many proofs I present, most of you are too lazy to accept that I have solved the riddle. Shown disinterest, childish belief in authority and indifference mixed with prejudice towards the development work of others, together with the desire to take a cheap ride on such work, provokes and is the explanation for the tone of the following three new pages.

A brief piece of advice, **before** you flip through the book.

If you are lazy and uninterested: Put it back!

Explanation to the lazy one, if that person is not too lazy to read a whole page:

Unfortunately, many people, like you, have the habit of judging a book without reading it. They flip and look at the pages in a skeptical and disinterested manner. They read a little here and a little there, saying they don't understand anything. **This is extremely disrespectful!** Do you also read a detective story, or a novel, like that? There is a reason why the pages of a book are written in a certain order.

The book is written so that people with **normal talent and without higher studies** will be able to understand it. One such reader, an old lumberjack, has even given me a valuable reflection, which is now also included in the book. But genuine and solid laziness kills any interest, when it is required that brain cells under a thick layer of dust, and in deep sleep, must be awakened. That's why it's better if you do something simpler, and do not insult the book with your presence and opinions without even having read it.

The book's theme, theoretical physics, is the most important science. The book has, thanks to common sense, made the subject less theoretical and more accessible. It also presents an explanation and proof of what gravity is. **Therefore, it is remarkable that you are uninterested in reading this book.**

Those who claim to be interested in science have often revealed themselves by being too lazy to read and absorb the logic of this book. Then it's more convenient for them to chew a gum, and uncritically read glossy popular science articles with few words and big letters. Then they can dream away and let themselves be seduced by lies in beautiful colours, and then pay tribute to the tailor, Einstein, who sewed the emperor's new clothes. Even though they don't see the clothes, they sing along to the chorus of praise. Why do they do that? Because it is convenient not to think for yourself, and thus keep the lie alive. That's how it's always been. If the high priests said they had to sacrifice children to appease the gods, they did so. If they said that women were witches and had to be burned at the stake, they believed that too. Had they lived in Galileo's or Darwin's time, they also would not have sacrificed one moment to understand the logical explanations of said gentlemen. Because it is more convenient to believe in an established lie from the dictatorship of the truth monopoly, than to think for yourself. They would cowardly and disinterestedly let the real truth-tellers burn at the stake. Unfortunately human DNA has not changed for the better. Most of you still believe every lie that comes from the monopoly, and are **too lazy and uninterested** to think for yourself and take a responsible stand. If everyone had been equally irresponsible as you, we would still be in the middle ages.

Only idiots praise other idiots, and allow themselves to be fascinated by idiotic claims such as one photon knowing what another photon is doing. Furthermore; Those who believe in the "Big Bang", and that it would also have started from our place in the universe, are no better than those who in the Middle Ages believed that the Earth was the centre of the universe.

Glossary: Lazy = When you refuse to do things that is outside your zone of interest.
Responsible = To act on what you see is wrong, even if it is outside your zone of interest.

You, who were not eliminated, can continue reading the book here.

The eternal struggle of truth against dictatorship.

"Question authorities!", Thor Heyerdahl taught us, and he had good reasons behind that statement. **Authorities, who consider themselves to have a monopoly on the truth, have often throughout history, been an obstacle to the development of science.**

Theoretical physics is theoretical. Therefore, the need for logic and common sense is extra great when analyzing the results of conducted experiments to check theories. But scientists do not automatically have a high IQ, just because they are scientists. And a possible correct conclusion does not mean everything from them is correct.

Theoretical physics is the most important science. The interaction between the strings of dark matter and the subatomic particles controls everything that happens, and lies at the bottom of all other sciences. The book has, thanks to common sense, made the subject less theoretical and more accessible. It also presents an explanation and proof of what gravity is.

Unfortunately, the wisest thoughts are not always valued at the compact wall of folly. One such opportunity is provided by Nobel laureate Roger Penrose. He believes, just as this book claims, that there has never been a Big Bang, which is not only a theory, it is also an illogical theory. But common sense is underrated. The book explains why it is illogical, even though it shouldn't be necessary to explain. Scientific journals and Swedish state funded television are referring to the Big Bang as something that has occurred, which should be prohibited. Spreading lies and unproven truth as truth is not "public service". Roger Penrose also mathematically described the existence of black holes. Einstein's attempt in that area failed, when it resulted in a description of something infinitely small.

The folly has ancient origins. When Einstein was piecing together his theories, making bricks without straw, he first had to construct two so-called postulates on which his theories could be built. One was that light has a constant speed and the other was the conjecture that we don't have just one frame of reference in the universe. But how can theories based on a conjecture, without source criticism, be accepted? Just because leading scientists had trouble explaining the result of an experiment conducted in 1887? These men were real scientists, and like this book, they did not accept Einstein's theories, but unfortunately they died of old age before the opportunist Einstein.

The claim that light has a constant speed is not true for extreme environments. Experiments with laser light through heavily cooled gas show a slowing down of the light. However, having multiple reference systems is the big and unforgivable figment. Read more about that later in this book.

But read everything in the right order, so that you first learn and understand the background to "The Role and Properties of Strings". Then you will also understand what actually happened in said experiment in 1887, as well as many other aspects of physics, such as; why the photon is not a particle, but a wave through dark matter. It would be irresponsible of me not to correct that misunderstanding.

At the beginning of the 20th century the old scientists were in the minority in relation to all scientifically uneducated and sensation searching journalists with their millions of naive readers. This mob was thus already in the majority, in relation to those who have tried to stand up for rhyme and reason. As we know, history has always been written by the victorious side. Today's rulers in the field will also ban this book. They will fight for their status and survival. Their Achilles' heel is that none of those, milking their sacred cow, Einstein, has so far found a critical fault in this book. But one must humbly realize that their arrogance and fear prevent them from reading it. As long as I cannot count you as supporters in the fight to end this shameful epoch in the history of theoretical physics, the cause is lost.

The future of the book...

The brainwashed individuals in theoretical physics have so far and will continue, as long as possible, stubbornly and arrogantly refuse to read this book. They will speak negatively about it and refuse to acknowledge its evidence, which they have not even read.

They see the book as blasphemy against their sacred cow. For that reason, they display a religious-like aggression, which is based on a fear that the world on which they have built their career will collapse. They will therefore not be impartial and fight for the truth, but instead they will fight for their status, income and self-esteem. Since, in short, they will be fighting for their lives in their own little world, they will never willingly admit the truth, but instead try to silence it as long as possible, and then devote themselves to smearing both the book and me. The mass media and the silent mob will habitually, for reasons of convenience, choose the side of the "high priests". Just as when Darwin tried to launch the theory of natural selection. Also Copernicus had a hard time when he tried to get farmers and priests to understand that it is the earth that moves around the sun, and not the other way around. The common people were uninterested, and the priests were afraid of their jobs. Man has not changed.

Hopefully it will be faster this time for the truth to prevail. And it will, even if today only a few stand up for it. The fact that I, among them, cannot count those in my immediate environment shows how non-existent the interest is. But "the combined weight of you all" will pave the way for reason.

This whole project, where I offer our world's greatest discovery/insight and possibilities for new vehicles, aircrafts included, and unlimited access to energy, for example via rotors with parts that are alternately weightless, has so far proven to be a waste. The discovery, which normally should have made me infinitely rich, is ignored because of, and with the help of, the established bastions of arrogance.

...in an indifferent and brainwashed present.

In a democracy where we can elect our government. But in theoretical physics, the researchers can dictatorially decide which theory should represent the truth. They rule over both state media and innocent people, just like in states we don't want to be compared to. And when state media claim, for example, that the Big Bang has occurred, the already brainwashed people don't protest.

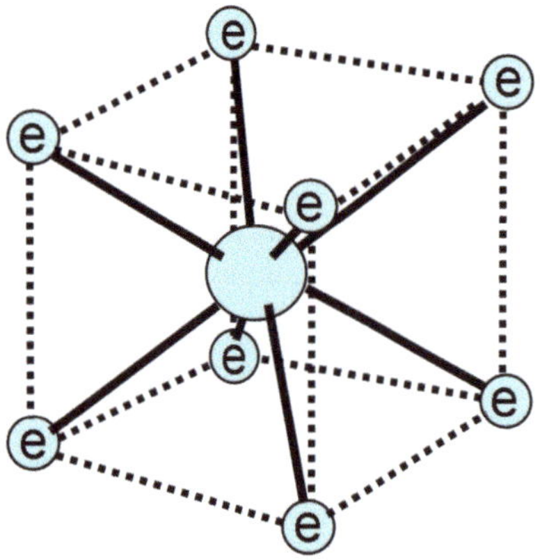

The figure shows an atom, which has 8 electrons in its outermost electron shell, and how these are connected to each other by the means of string chains (dashed lines), which are formed by strings from the dark matter.

An atom's outermost electron shell can also consist of a number other than 8 electrons. But the fact that an atom, which has an outermost shell of 8 electrons, cannot bond with other atoms tells us that each electron has 3 string handles, no more no less.

With these 3 string handles, 4 electrons could also form the corners of a tetrahedron (pyramid) in an outermost shell. But such an atom apparently becomes more exposed to the environment, probably because the attachments of the string chains to the electron become too weak at smaller angles between the string chains, possibly due to the handle's location on the electron. The cubic form is the smallest possible symmetrical form and thus the strongest, which can be formed by eight electrons. With a number other than 8 electrons in the outermost shell, the atom, as understood from the above, gets free string handles towards the surroundings, and can then form molecules.

The electrons inside the depicted shell are not in the picture, but those of their string handles, which are not coupled internally, couple to the electrons of surrounding atoms. This is what makes up both gravity and the Van der Waals force.

As simple as that!

The background to the following conclusions is based on the book's content.

1. The electron has three handles, which by the means of "string chains" of dark matter, connect them to each other. A fourth and special handle connects the electron via another type of string chain (solid line in the figure above) to a proton, in the nucleus of the atom.

2. The three aforementioned handles explain why;

a) a noble gas, with its 8 electrons in its outermost shell, is a stable element, because it has no free string handle in the outermost shell that it can connect to other atoms, and therefore remains in gaseous form, even if it would weigh more than lead.
b) atoms in other elements form stable molecules, to achieve 8 electrons in the outermost shell.

3. Free handles from an atom's electrons connect to the surroundings and form the gravity acting on the atom.

4. The electrons in helium II, which only exist in experiments near absolute zero in laboratories, have no free handle from the electrons. Helium II therefore becomes weightless, which proves the above conclusions, namely that the connections of the handles to the atom's surroundings constitute gravity.

5. Shows a He_2 molecule (helium II) with two double bonds between electrons, and subsequently without any available string handles, for "mediating" gravity. Even in those cases where the molecule has only one or no double bindings, it still has unusually few available string handles, which should mean a lower weight than the combined weight of two separate helium atoms.

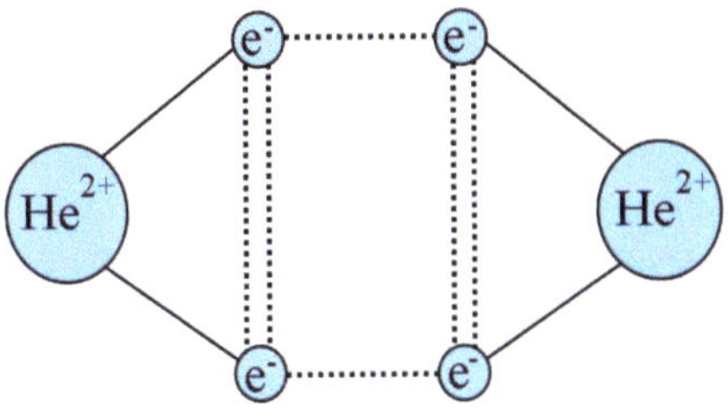

Important!

1. With the description, provided in this book, on string chains that exists between subatomic particles as the basis, it could **first** be stated that a molecule consisting of 2 helium atoms, if it exists, should be almost weightless.

2. Descriptions of helium II was **subsequently** found on the internet. See the following web site (available in 2014): **http://www.youtube.com/watch?v=2Z6UJbwxBZI** Observe the expression **"It appears to defy gravity!"** in the video.

3. The web site is a proof that the **theories in this book is an accurate description of reality**!

String chains as a concrete reality is proved by the frog that lay on these, as in a hammock, in a very strong magnetic field. This was described in Section 1, under “"Magnetic" repulsion does not exist”

Thus, the search for weightlessness has been successful. This knowledge is the first step in the development of a new type of vehicles (everything from cars to spaceships) and energy production, both categories with unimagined possibilities.

Welcome into the dark matter!

Now remember that you do not belong to the category that falls into leafing through the book tiredly, only to put it aside! The next two pages are a great introduction.

The science of particle physics holds all the other branches of science.

Matter versus so-called dark matter.

The book argues that so-called strings in the dark matter constitute the building material to both chains of strings and subatomic particles envisioned as strings that cling to each other. Different strings and their varying combinations result in different outcome. An insight and comprehension that there is a constant ongoing interaction between particles, strings, and chains of strings will help the reader to understand the conclusions of this book.

Once you have read the book, you will understand how chains of strings mediate and cause gravitation. The book explains also how they are not affected by the gravitational force despite their mass. It was this detail that was not understood, leading the scientists to adopt a wrong tract at the end of the 1800s. Even though the atomic architecture was not yet identified, the ideas from that time are still allowed to influence modern particle physics. And even though all branches of science merge at the most fundamental level, possible chemical aspects within the particle-physical phenomena have presumably not been considered, and questions, such as why do atoms "wish" to have 8 electrons in their outermost shells, have not been raised by particle physicists. This book, however, both raises the question and answers it.

The book is controversial, yet it supports all its claims and conclusions on data established by recognized institutions, which should stop possible criticism.

All educational institutions feel obligated to guard established "truth", even old unrealistic figments. For example, following the publication of their theories on evolution, it took decades for Darwin and Wallace to get them accepted. They had the logic on their side, but against them, they had the well-established Creation.

This book is likewise based on logic. However, the history has shown that neither evidence or proof is sufficient to overcome established concepts in just one generation. But the hope is that you read this book with an open mind. Together we can make a difference, and together we can challenge the poor odds dealt by the history.

The figure illustrates how the book describes electrons as conjoint strings coupled to each other with a string chain, and how this is challenged by surrounding string chains and single strings.

– And yet, it (the Earth) moves.
Galileo Galilei

Close to four centuries later...
– And yet, the photon is not.
Rolf Sjöström

To gain the knowledge presented in this book, no special "prior knowledge" is needed, just a keen interest.

Explanations in this book are logical and **easier** to understand than the established falsehoods about so called photons and changing speed of time.

Everything consists of something

This applies also to the smallest measurable particles in an atom. Therefore, it is logical to assume that they are build up by even smaller particles that are below the detection level. Equally logical is to assume that these particles compose the so called dark matter, existence of which is both accepted and confirmed by established institutions today. On this foundation, the following claims can be constructed: Particles (so called "strings") in dark matter behave like small chain links, sticking to each other, building chains of strings and even detectable particles. These – for example a string chain and an electron – can further get latched together (see the figure on page 1) to then break apart by outer forces, which forms the base to explain everything within particle physics and chemistry.

Based on the above, the logic will further us to assume;
1. that there is a continuous interaction between dark matter and (visible) matter, and
2. that movements passing through dark matter are part of this interaction, and
3. that these movements should expand like rings on water though the dark matter and result in effects that have been incorrectly accredited to the so-called photons. As evident from the contents of the book, referencing scientifically approved experiments, the above assumptions align so well that they rise to be more legitimate than the photon.

Just because Einstein's theories are still alive, they do not need to be correct

On the other hand, it means that **they work** for waves formed by strings **even though** the theories are claimed to apply to the so-called photons. And where these wave formations change direction due to changes in string density around a star, gravitational force is used to explained the effect on these so-called photons. "Survival" of Einstein is also based on people's need to adore and praise gods with their "infallible" truth. Therefore, those scientists who should read this book will never open it. The have already painted themselves in a corner with their photons from where, unashamed, they now try to find a way out. History shows also that neither reason not shame has ever fit in places filled with preconceived ideas or blind media idolization of authoritarian belief systems.

To be rooted on a stable foundation, cultural literacy should rest on knowledge about the origin of all being.

The unfortunate damage,

inflicted by Einstein on our understanding of the universe was so extensive that a belief on so-called ether (dark matter) with its strings and string chains became "illicit". The blame belongs primarily to the fantasy-filled explanations by Lorentz on an experiment run by Michelson and Morley in 1887. Stokes' and Planck's theory that matter pulls the ether (strings – string chains) was disapproved by Lorentz using mathematics. However, to disapprove a theory, more is needed than a mathematical calculation that does not fit in. An incorrect assumption can indeed result in selecting a wrong mathematical model.

To better understand the book, read this part first.

Basic information about the atomic structure
A repetition from school books and other available knowledge

The figure below depicts an atom, specifically a carbon atom. Colors do not exist at this level but to demonstrate the different subatomic particles we have designated specific colors to them. The atomic nucleus consists of the "red" protons and "green" neutrons. The electrons, shown here in blue color, can leave an atom easier and subsequently they mediate electrical current, as it is called. All atoms of chemical elements are build this way; the number of protons define the element, and their weight is defined by the number of protons and neutrons. The traditional view is that the electrons orbit the atomic nucleus. This view, however, is not fortified in this book.

Protons and electrons in an atom have been shown to attract each other. This attractive force that holds the atom together, is defined as electrical potential where electrons are defined as having negative and protons as positive charge. When an atom has one less electron than protons, this atom will catch the first available electron. Likewise, and in line with the above, if an atom has more electrons than protons, the "unbound" electrons can carry electric current.

The neutrons do not share this capability, defined here as electric potential, but without neutrons, an atomic nucleus with more than one proton cannot keep itself together. Normally, an atomic nucleus has equal number of protons and neutrons. The size of an atomic nucleus is limited, and there is a limit for how many more neutrons than protons an atomic nucleus can have. Once surpassed, the atomic nucleus becomes unstable. The disintegration results in emission of so called radioactive radiation. Beta ray indicates that a neutron is a proton plus an electron because a neutron, upon emitting an electron, is converted to a proton. (In alpha decay, so-called alpha particles, 2 protons + 2 neutrons, are catapulted. Similar atomic nuclei colliding with each other initiate chain reactions and cause high-energy gamma radiation.) The outcome of radioactive decay is other chemical elements, naturally.

What has also been shown is that an atom with 8 electrons at the outermost shell does not fuse with other atoms, and that molecules, such as water molecules, are formed when the atoms together can get a shell with 8 electrons. All this basic information about the atomic structure forms the essential cornerstones used to construct the conclusions of this book.

Figure depicting an atom. This is a carbon atom.

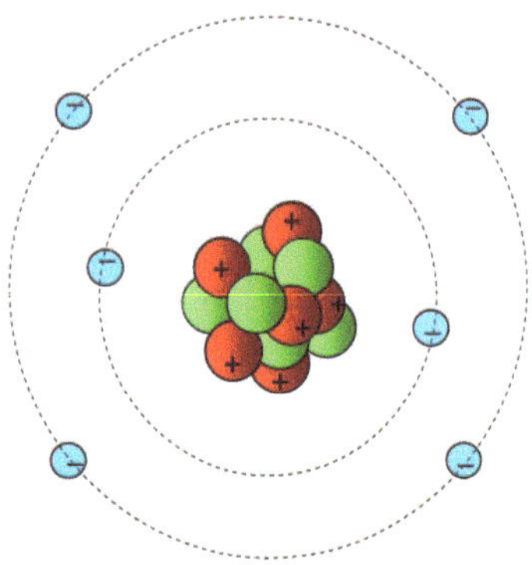

Introductory summary for greater understanding of the book

The book is not a simple novel; it is about particle physics, a topic that is usually not discussed at a kitchen table. And for scientist within the filed, the book entails a fundamental change in concepts making it especially difficult to accept despite all the references to experiments from recognized institutions, accepted facts, and subsequent logical conclusions. (Other than conclusions are described in the book as logical assumptions/hypotheses, and nothing else.)
The below items, both introduction and summary, have been written because it is always easier to accept what you will meet if you are prepared to what lies ahead. Due to the brief character of the items, they do not contain all the logical reasoning and references; these are provided later in the book. Therefore, they cannot be criticized for not being thoroughly researched. Anyone wanting to criticize the book should first point out factual issues inside the book. Similarly, this critic should also point out how today's recognized particle physics could be considered more credible when an outcome of a single scientific experiment on light forces today's "faithful" scientists to conclude that so-called photons are aware of each other's fate!
(See: **http://en.wikipedia.org/wiki/Delayed_choice_quantum_eraser**)

1. Most people accept that the universe consists mostly of so-called dark matter. The book refers this dark matter as strings. The book argues that these strings can stick to each other and form chains and subatomic particles. This argument is based on the logic that everything – including the subatomic particles – consists of something smaller in size.

The culmination of the conclusion that this logical assumption leads to is described in Item 12, referring to an experiment published online. It describes helium II behaving just as the undersigned had expected, that is, without gravitation. Noteworthy is that those who performed the experiment have no explanation to provide in the video to helium II's weightlessness.

2. The subatomic particles that are formed as described above include electrons (e). And since they, as described above, consist of strings, they can be assumed to have a surface that allows them to latch to adjacent string chains. It is this assumption that is central to what comes next.

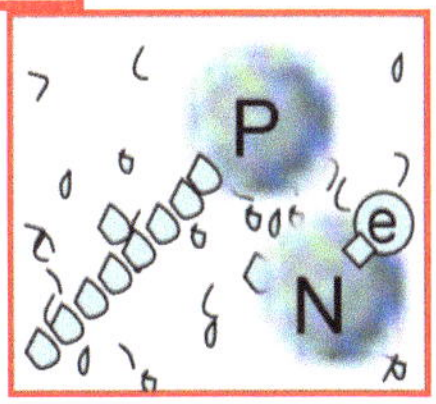

3. Also protons (P) are being formed. Entire skies of ionized hydrogen gas (= protons) have been observed by astronomers. And since the protons consist of strings, they should have a surface that allows them to latch to a sting chain. From the book's logical reasoning demonstrates that this string chain could be what is usually referred to as positron. How a proton's 3 quarks are turned to hold an attached electron, and become a neuron (N), is considered a key to this question. Protons and neutrons are called nucleons.

For those with a very keen interest: These quarks are thought to be turned in one way in relation to each other in a neutron, and in another in a proton. The book argues that this explains how an electron or positron can get stuck, and that a nucleon with 3 quarks that do not hold a positron or an electron, is a particle that is called for "otron" in the book. An otron is like a particle; by definition, it is dark matter because it does not constitute a part of "our" matter.
Atoms with more than 1 proton in their atomic nuclei contain about the same number of neutrons and protons. Likely with the help of their electrons, neutrons work like inter-proton glue in the atomic nuclei.

4. We know that a hydrogen atom consists of one proton and one electron. The accepted theory of how an atom is coupled to a proton, is based on the concept of attraction. However, an attraction must have a medium that mediates the force. The book adopts an assumption that this attraction consists of a coupling between the electron and the string chain of the proton. And it is when an electron gets latched on a proton's string chain/positron when a hydrogen atom is formed.

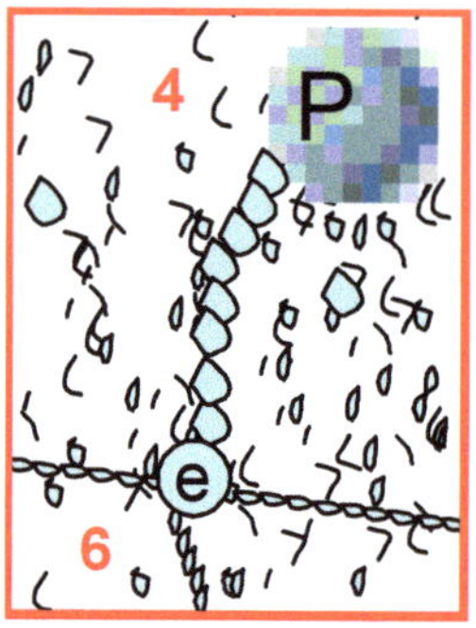

5. The book explains the attraction and its opposite force, repulsion, with strings that are moving due to a constant interaction with their surroundings, and when they collide, they either get latched to each other (attraction) or they just bounce off each other (repulsion).

6. The book argues also that in addition to its linkage to a proton, the electron has handles facing other string chains. Through this type of string chains, the hydrogen atoms' electrons interconnect in the universe to form large clouds of hydrogen gas which have also been observed by astronomers.

7. The clouds of hydrogen gas keep growing. Finally, the amount of intercoupled forces becomes so large that the compressed hydrogen atoms are merged to form other and larger atoms. A star has been born. Moreover, strings cannot be compressed by gravitation because the strings are those forming the "ropes" that draw matter together with help of the force we call gravitational force (or Van der Waals force at the atomic level).

This process of spontaneous creation of subatomic particles and hydrogen atoms is expected to continue interrupted and everywhere, from hydrogen gas feeding the sun to hydrogen gas that is being created, among other places, inside the human body. Availability of strings that are the building blocks for subatomic particles especially must be the limiting factor.

8. Helium atoms are formed when hydrogen atoms are compressed. (Four hydrogen atoms are needed to form one helium atom.) Our star, the sun, is at this stage. The stars mature further by pressing the resulting atoms into larger and larger atoms, also those with more protons and neutrons in the atomic nucleus, and with the equal number of electrons and protons. This is how all chemical elements are being and have been created.

9. Atoms with 8 electrons on their outermost shell (inert gases) display a characteristic of not being able to attach to other atoms. Further, if we imagine such atoms having a form of a cube with 8 corners, and that its corners (electrons) are tied to each other through the edges (string chains) of the cube, we must realize that <u>each electron has exactly 3 handles reaching towards string chains that in turn latch themselves to other electrons</u>. This forms each electron's linkage to "its" proton in the atomic nucleus, mediated through its dedicated string chain (positron). The left figure shows an atom for the inert gas neon.

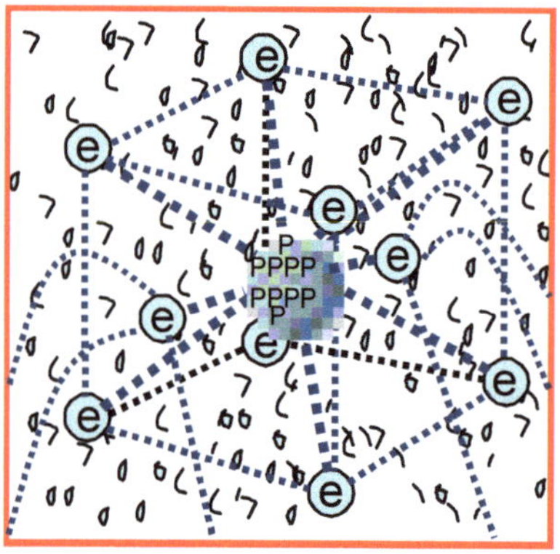

10. An atom with 7 electrons on its outermost shell has thus free handles/string chains that can hold an 8th electron that belongs to another atom to form a molecule. Table salt molecule, sodium chloride, is an example of such a union. The right figure depicts an oxygen molecule with an outermost shell with 6 electrons that has "borrowed" 2 electrons from 2 hydrogen gas atoms to form a water molecule.

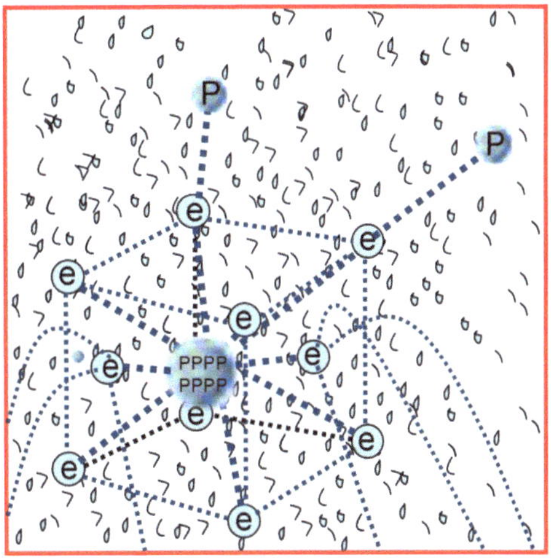

For those with a very keen interest: A string chain already bound to the electrons of the 2 hydrogen atoms, could explain why these hydrogen atoms are located on the same side. Another possible explanation is that the 2 available corners of a "free" oxygen atom are presumably already located on the same side because then its 6 outer electrons will have a maximum number of linkages. Electrons in hydrogen atoms are not completely attached and therefore the water molecules become polar, which explains why they latch to each other and to salt ions (divided salt molecules), if present.

And as depicted in this and the above figures, the inner electrons of these atoms that resign on their own so-called shell have the available string handles facing the surrounding. The inner shell may be explained by the geometry of an atomic nucleus that could prevent these electrons from having secure linkages to the electrons on the outer shell. However, the book does not stand or fall by this. (The concept "theoretical physics" reflects its real significance here.)

11. In other words: The available string handles latch themselves to adjacent string chains that can be of unlimited length. This explains gravitation. Inert gases, and the molecule described above, could be in a state of zero gravity (weightless) if it was not for the fact that electrons inside the outermost shell carry available string handles to form linkages with the surrounding.

12. Helium II is a volatile fusion between 2 helium atoms at a temperature close to 0 degrees Kelvin. Helium II has weightless properties, and by inspecting how these atoms are coupled together it becomes obvious that their electrons present only a few available string handles that can contribute to the gravitational force.
If we then observe another phenomenon seen in superconductors at temperatures close to absolute zero, namely the formation of a so-called Cooper pair by electron linkage, this should equal to double inter-electronic string chains in helium II. When this takes place, there are no more electrons with available string handles in this molecule which now **gains zero gravity**, also shown in the experiment referred to in the book. Even in the absence of the Cooper pair, the weight of helium II would be half of the total weight of the helium atoms, based on the number of available string handles.

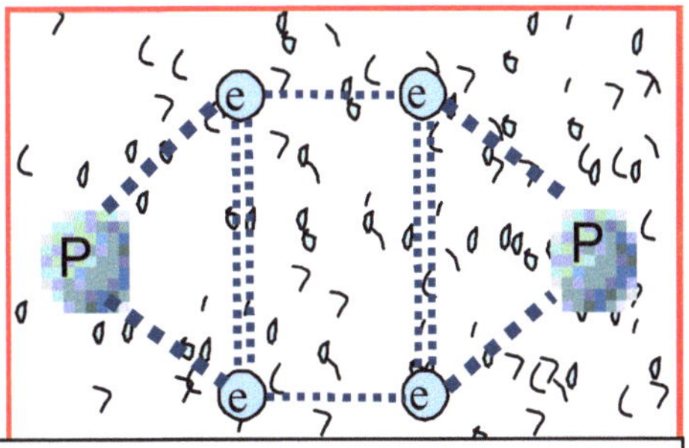

See also the following web site (available in 2014):
http://www.youtube.com/watch?v=2Z6UJbwxBZI
Observe the expression you hear in the video:
"It appears to defy gravity!".

13. The above argument does not hold the whole truth. The neutrons in the atomic nucleus contribute as much mass as the protons do. Moreover, the out-facing string handles of the “build-in” electrons of the neutrons add to the atomic weight. A hydrogen atom with only one proton in its nucleus doubles thus its mass and weight with one neutron. This is not as simple for the remaining of the atoms but an atom’s weight should provide a clue how protons and neutrons are placed in the atom's nucleus. A neutron in the middle of a larger atomic nucleus, for example, should be unable to form linkages with its surroundings, and for the above helium II atom, “internal” linkages can be thought to involve neutrons.

14. The book shows also that what was described incorrectly as photons (particles) for a period of about 100 years, are waves crossing the ocean of strings/string chains that surround us. The book refers also here to accepted experiments. The so-called photons are said to have mass/energy only when they move. A wave on a water surface must also move to have mass/energy. A wave movement, like the light, has also a constant speed through a given medium. It can be mentioned here that gravitational waves should be the waves that follow the gravitation-mediating string chains. This would explain why they are said to have a higher speed than the light as these waves are carried along the strings that are linked together.

15. The claim that the speed of time for an object would change along its speed is the greatest, and unfortunately the most successful, hoax of the century. It originated from Lorentz’s interpretation on an experiment performed 1887. “Ether” (the strings) was known at that time but precisely like today, scientists were not able to formulate the ‘Role and characteristics of the strings’, as they are described in the book. The scientists were apologized this time because they did not know the atomic structure. The later epoch during which the ether (the strings or dark matter) was denied, did not make it any easier. Therefore, these experiments could not be explained using logic. And with no logic available, something so illogical as an assumption that the speed of time for an object varies with its speed, had to be adopted. This leads to a logical objection and question: Speed in relation to what?

If we could show today that an atomic clock has slowed down (apart from the clock’s own lack of precision) after a few rounds around the earth, we would have not considered how interaction of the subatomic particles should be affected by the changed circumstances, such as changed density of the string chains that mediate gravitational force, among others.

16. Closely related phenomena, such as electricity, magnetism and (indeed) kinetics are discussed in the book's Annex. Black holes, and other concepts that belong to the universe, are discussed to complete the book. We will also get to understand why we can forget about the imaginary spectacle “Big Bang” and to adopt the established concept of “Steady State”. Even superconductivity and superfluidity are described and explained because everything is interconnected.

Those assigned to guard the science, keep up with their earlier colleagues who for centuries clutched to the notion that everything is composed of earth, fire, air and water, despite the discovery of metals, such as copper and iron. Our so-called scientists have – paradoxically enough – always been the greatest hinder for fundamental changes within the science. Just like religion, understandably, the established science must be kept orthodox. This is now, however, occurring at the expense of logic.
And while fighting such a dark establishment, we may waste centuries instead of use the time to develop particle physics. The number of proof and evidence supporting this book has no bearing. Hence, Thor Heyerdahl’s words “Question authority!” have merit.

The Book

A brief presentation of the contents

Section 1. To lighten up the tone above the dryness typical to scientific dissertations, the book's preface was given a personal touch. Next, the puzzle will be solved by using pieces of already accepted facts. The pieces used in the book to complete the puzzle with, are the details of the role played by dark matter. Step by step, using pure logic, the puzzle will be solved. Section 1 is concluded with 'Role and characteristics of the strings', a conclusion never before formulated.

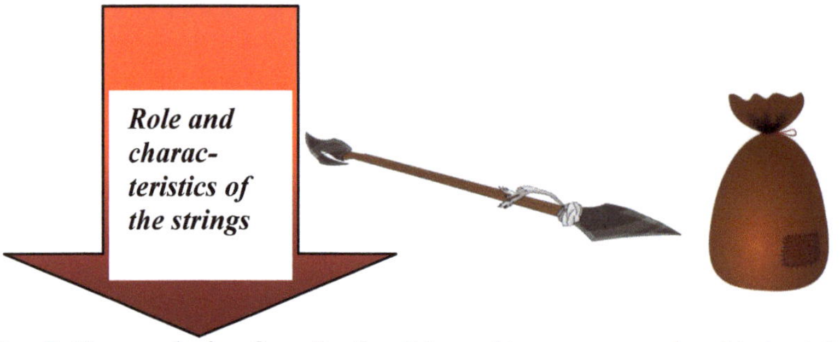

The sack with fine mathematics bursts to expose its core, Lorentz factor, born as a result of an incorrect interpretation of a 1887 experiment.

Section 2. The conclusion from Section 2 is used here as a spearhead to bust the myth of the variable speed of time. The myth originates from an experiment performed in 1887. At that time, the ether (dark matter) was accepted but the atomic structure was still unknown, and hence, it was not possible to come up with 'Role and characteristics of the strings'. When the problem was later solved by assigning the time a variable, development within particle physics got off the rail. Beautiful mathematical formulas written to describe this delusion did not make the matter better. It went so far that the concept of light as an undulating wave was replaced by light as particles. This has led to other delusions, such as a photon knows what its twin photon does, according to an experiment. As shown in Section 1, a photon is not needed to explain why a helium atom is lighter than four hydrogen atoms which are merged to helium upon emission of radiation.

Section 3. Proves the accuracy of the theory by means of helium II's characteristics.

Section 4. In this section, the theory is applied to kinetic energy including an explanation to why this energy is increased by the speed squared, which should not be the case because the speed is always relative. The predominant theories on neutrinos, and the Casimir effect, are marked here with a question mark.

Section 5. General information about the universe we are a part of, and a few philosophical reflections. All in the light of what the book has explained so far. And who for example has heard of "Steady State"? Because the book nudges at the unexplainable, it will be ended with a few points touching philosophy and religion.

An interested trip together with the author is offered to those who choose to read this book This has been a trip with new conclusions drawn, and that became increasingly more ambitious as the image got clearer exposing the connections. Important; The contents of this book are built on pure logic, and solidly anchored with references to discoveries and experiments carried out by recognized institutions.

Section 1

Subatomic Particles and Their String Handles

Proton plate, or the art of nullifying gravity

The reader is reminded here about the atomic structure and about the forces identified inside and around an atom. By summarizing these known facts, both the undersigned and the reader will be made ready to continue with the book.

Preface – A personal and historical background

During an early spring of 1974, a 2.5-year long period of thinking and preparations culminated in an experiment. All the thinking and preparation had been going on parallel with my daily life, first as a student in Uppsala, followed by military service in Söderhamn, and last as a student in Sundsvall.

It was dark outside, and the curtains covering the window in my tiny room were closed. The electro-static generator was running at full speed throwing out 2-inch-long lightning sparks while I attempted to lead the high voltage through a golden electrode in a glass tube containing hydrochloric acid. The experiment took place in a small room I was renting since the fall semester. My two landladies had a white-painted house located on 22 Branta vägen (Steep Road). Above the staircase next to the front door, they had painted "Tubo" (Nest for two) with blue letters. They lived on the first floor, and rented out a couple of room on the upper floor. They were already rented, and I got to rent a room at the first floor for 170 SEK a month. The room was at the end of a hall way on the right, and made out the northwest part of the house. The kitchen was located on the left end of the hallway.

My desk was below the only window, facing west, and the bed along the north wall. A chest was right of the door to the room, and a green flat bowl with water was on the chest. Dismay and horror filled Ida and Beda – these were in fact the names of my landladies – when they realized that the puddle in the bowl had come to life. I did find it fascinating myself. "Sea-Monkeys" was written on the small bag containing the life-giving tiny seeds. My landladies were now on high alert on what my experiments could cause for a mess, and dropped by every now and then. Naturally, I did not dare revealing my ambitious plans. It was bad enough already. Everything was rigged and ready on my desk. Before reaching this long-awaited milestone, there had been a heap of preparations and a lot of hardship. My great plan had cost me my social life and parts of my student loan. Occasionally, I threw a jealous eye on my sea-monkeys already ready to go for it with a mate. They should at least thank me. Well, once I have managed to climb out of the hole, it may be my turn, I thought.

It all began at fall 1971 in Uppsala during my first year away from home, when I came to think about the similarities between gravity and magnetism. What is magnetism? We know what is needed to generate magnetism, and at the outset, the phenomenon can be explained with electron currents. What is gravity? Interchange of the not yet confirmed protons? We know that gravity exists between all matter, down to the atomic level. But where is the gravity in a dissembled atom? What I did not know was that Einstein, until the very end, until his passing 16 years earlier, had tried to solve the mystery of gravitational force that is considered one of the four elementary forces at the atomic level. I myself begin to lean toward thinking that it was not an independent force but rather it behaved more like magnetism at the atomic level. Maybe the solution was not so very far, and why search a depth that may not exist? A lot of the time intended to study advanced mathematics this fall was instead used to draw, and to wind copper wire around reels in preparation for the upcoming experiment with magnetism.

Military service in the following year put a stop for any mental exercises furthering the experiment. Instead, a strap, an extra strap designed for battle gear became my contribution for human civilization that year. So, when fall 1973 was here and I was back at my task to unravel the mystery of gravity, I was equipped with a rather soft mind after the nice mental vacation at the Air Force.

The best-known force in an atom is the force between a proton and an electron, the force that gives rise to electricity. Because the protons in an atomic nucleus should have an attracting effect on other atoms' electrons causing thus atoms to be drawn toward each other, it is natural to assume that same forces can explain gravitation. But how could a group of protons, without the surrounding electron cloud, react on the surrounding world if they are isolated in a glass container. Should its repelling force on the protons of the surrounding atoms be greater that the attracting force on the electrons of the surrounding atoms? I drew and I thought. Among other things, I took into consideration the wider distribution of the electrons, and subsequently their more fragmented attractional force.

It is time to seek answers through experimental methods. I ordered an electrostatic generator from Norstedt in Stockholm – my largest investment. And in an ice-cold winter evening, I walked out from a small Bernina store in downtown Sundsvall with sewing machine engine I got for 25 Skr. Next, a piece of gold from a jewelry store that I hammered into a nail shaped electrode. A pipette that was delivered together with the electrostatic generator needed to form the pipette into a desired form with a blowtorch, all with my brother's help: first, the pipet was cut to a length of 10-15 cm, then one end was bent up and last the other end was drawn to a thin tip with a microscopic opening.

This was the modus operandi for the experiment where the plan was to achieve electrolysis using a chemically stable anode-electrode -the gold – causing a larger excess of protons to remain in the glass tube. Chlorine from the hydrochloric acid would also be escaping past the gold. The protons would accumulate in the sealed end with the help of magnetic field from a copper coil wound around the other end of the glass tube. Maybe I also had a field of negative electrical potential on this side generated by the electrostatic generator. Weakness of this design was the separation of hydrogen ions, i.e., the extraction of protons, that had to be driven by electrolysis without a cathode, but compensated by a large electrical potential.

With mixed emotions, I started the sewing machine engine, and let the electrostatic generator thunder away. I was equally worried that something would go askew, or that nothing would happen at all.

Nothing happened. At least nothing that could be registered. I had been feeding a small hope of something, possibly a smell of chloride, but not even that. Well, both Ida and Beta could relax, this time. And yet, I was satisfied; I had given it a try. And I had, possible, inhaled a few lost protons because a corresponding number of chloride atoms would not have been sufficient to notice.

Had the expected chloride smell been clear, the next step would have been to extend the tube with a copper coil and connect it to a large flat glass container that would have become the largest proton plate in the world of its time. Even the concept "proton plate" would have been novel. At this stage, if ever reached, the Greek Hermes' magical seal would have been needed to seal the glass tube and the crystal structure of glass to prevent electrons to seek their way to the desirable protons. I spent time laying on my bed thinking. Maybe I should have made the "proton end' of the glass tube longer? Was the magnetic field from the coil in right orientation, and was the voltage going through the coil strong enough?

I started to realize that it was time to abandon my equipment, catch the life's over-crowded raft, passively float downstream the river, and like the most people, begin to pursue the most primitive human needs. Not a bad life, in fact.

In the autumn of '73, terrible scenes had been shown on the TV in Ida and Beda's living room. It was a brutal military coup in Chile, and at the Swedish embassy in Santiago, among many other, a young family sought protection. The daydream of the ultimate weapon had therefore inspired my attempts that winter. But now, the following spring, I was reluctantly forced to pack my laboratory. I got my first employment as a computer programmer three years later in Stockholm. In the same office, the mother of said family, also began working. We became good friends. A few years later I worked as a consultant in Toronto. After the end of that contract I moved down to Rio de Janeiro, in an attempt to clarify what had happened to my Brazilian wife and unborn child. One day there, in Rio, I was visited by a man who wondered if I could help his daughter keep her Swedish language alive. After the second lesson, I explained to the girl that her Swedish was probably better than mine, gave her a magazine with well-written works, and told her as a farewell that the next step for her was to see how journalists write, and that maybe she also could become one. Three decades later and back in Sweden, I recognized her, now as a well-established TV journalist. In one morning show, her guest was a physicist from Umeå. He was upset about the "deceiving" facts that now was available. It looked like he was holding an earlier edition of this book in his hand. Several years later, I discovered that this TV journalist also was the daughter of said co-worker in Stockholm. A circle was closed.

But before that discovery**;** After Rio two decades had disappeared in Norway. And finally back in Sweden, I soon ended up wasting additional time, again following derailing needs. This time by building a house for my next ex-wife to be, in Thailand. When the house was finished, I had lost the last kilo my body could spare, and been compensated with pain from two unforgettable scorpions. However; I now had, inside my head, the sequel to this book ready. Due to the difficult language, I had had the opportunity to think undisturbed. So after a long detour, for me and for you readers, I was finally back to where I left off, now armed with bright thoughts about dark matter and its so-called strings. The year was 2005.

Current status, summary

I have run aground a few times, only to learn that ideas are hard to turn into money. So, this report is sort of a delegation document of the proton plate. **Re-analysis of what was relevant to the topic and what is considered the convention** was thus needed. The outcome of this, and what the logic subsequently leads to, had made this publication ever more urgent. Therefore, I begin this delegation with the following conventional facts, described at a straightforward, correct and expedient level. Diving into “pions”, “antineutrinos” and other particles at this stage would be like abandoning a highway for a hike on a trail in jungle.

1. Protons have a strong repulsive force. This is usually depicted with a graft showing how it increases with decreasing distance before it suddenly changes to attraction.

2. For more than one protons to co-exist, at least one neutron is needed. Neutrons seem to function as glue. As an example, we can mention the alpha particle. It consists of two protons plus two neutrons. These four “nucleons” form a stable cluster which can be explained by one of the neutrons emitting an electron and turning into a proton. Consequently, like glue in the atomic nucleus, this electron should be able to move between protons that take turns to be a neutron.

3. Despite the attraction between protons and electrons, the nuclear protons do not accept electrons that surround the atomic nucleus to then become neutrons. (The exception: when such transformation leads to stabilization of the atomic nucleus.)

How the latter works, one can only guess. Maybe a proton is expelled from the atomic nucleus and is immediately hit by an electron from the closest surrounding shell. Item 2 would explain how it is then drawn back to the atomic nucleus. Why a free electron is not accepted with open arms by the atomic nucleus may be explained by a kinetic pattern that is different for an electron than for a neutron, or maybe by a linkage of the electron to something. I will return later in the book to postulate about this "something" including an explanation to why electrons are interested in protons.

Hence, the following applies:
A. Protons seem to behave repulsive towards other protons and electrons.
B. Despite this, protons are attracted by electrons. (A paradox.)

And the following seems to apply:
C. What seems to keep the protons together in an atomic nucleus is the electron shell that surrounds them, and that also neutralizes the protons' seemingly repulsive force toward their surroundings.

Gravitation could be explained by B, while A could indicate that I was on a correct path with my initial experiment, even if the level of ambition can be questioned. The logic of the analysis detailed on the following pages will be built on the three above claims. The result at the end of this section (Section 1) will be eye-opening.

In the late winter 1974, the thought was that the repulsive force from a possible proton plate would be controlled by twisting it in varying angles towards the earth. To screen it, using plates or a cover that partially transformed it to a capacitor, would be another way to control it. The major task was still to be completed. That is, to store a larger number of separated protons. The technique to separate protons and to send them further has existed for many decades. But is it possible to store enough of them without them attracting electrons? And if so, would the result be interesting?

Figure 3a. The modified pipette.

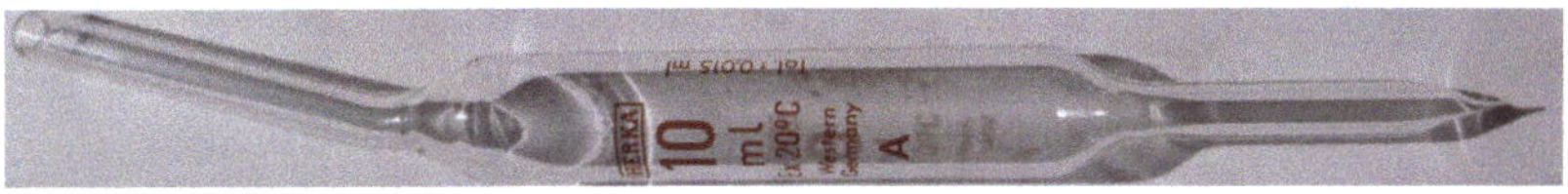

Figure 3b. Shows a side view of a glass plate with a cavity; The proton plate that never happened.

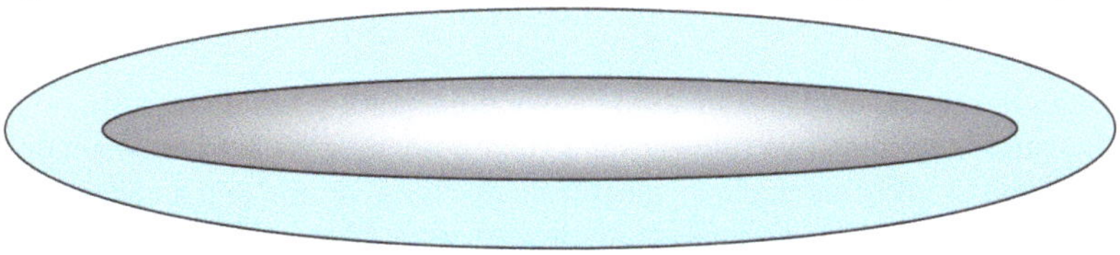

Advancing closer to Otron plate, the next-generation Proton plate

With the help of logic and established facts (which is consistent throughout this book) we show here step-by-step how strings-string chains are an inevitable component of an atomic structure. We also describe here the effects of the inter-atomic strings. Evidence to prove these theories is provided in Section 3.

The time is here to attempt to complete my self-imposed mission. It was 4 decades ago I reluctantly placed my experimental equipment into a storage box. Yet the ideas, too big to fit in that box, have been very much present through all these years. This time, though, I will rely on the logic that is based on the existence of the so-called strings.

It is well established that an answer always opens new questions. An answer is like steps in a staircase you are slowly climbing in your attempt to reach the next step that you are constructing using the logic and experience. Same applies to studies of the smaller components in the matter. Theories and truths can be anchored in the many steps to create a steady footing to climb on. As an example, we accept protons and electrons as a simple and operational explanation to electricity. We have defined them as different potentials that are striving to offset one another. To take a next step, you can, as always, ask "Why"? And what is electric potential?

Section 5 describes our universe as an endless ocean of strings (dark matter), and that the subatomic particles are composed of different conjoint strings. A belief that they would not be composed of anything is illogical. This is what we can use as building material for the first step, to then move forward to:

Subatomic handles, a new concept

There is no reason to believe that the said conjoint strings – the subatomic particles – would always present themselves as smooth small bolls without any protruding ends reaching towards one of the inherent strings. Assume, for example, that an electron has at least on such protruding string end, a "string handle". And since the strings can latch to each other and form the said conjoint strings, it means that any strings surrounding the electron can get stuck in the handle. Therefore, interconnections between subatomic particles and strings should exist. We already know that forces that hold together the inter-atomic parts exist. Hence, we can advance to build the next step based on these handles.

The types of subatomic handles

We begin by studying which handles we have inside an atom, above the "quark level" (a nucleon is composed of three quarks). Because more than one rule is used to compose an atom, we can assume that there are different types of handles. When a nucleon, a neutron, decays, a proton and an electron are formed. A looser association between these two makes a hydrogen atom. Consequently, the latter construction must be explained by an existence of a string or a chain of strings (sting chain) between the proton and electron. This string/string chain could be the positron that turns a nucleon to a proton.

In other words, what has been described as splitting of the neutron occurs because a positron eliminates the electron, and consequently, the neutron does not contain any proton. (See figure 12 in Section 4.) Two protons do not bind directly to each other because they have similar handles, or because they are too heavy for a string chain to hold them together. A proton looks different from an electron hence offering a possibility that the above linkage to a hydrogen atom could be described as a so-called female and male coupling. Later, we will see inter-electron linkages.

In an atomic nucleus, there is never more than one proton unaccompanied by neutrons. Two protons cannot be held together without neutrons. This constellation is exactly what is found in helium's atomic nucleus. This is where the assumption of handles fits well in. Because, while each of the two protons are linked to their respective electrons located outside the atomic nucleus, there is no handle to link a proton to another proton. Consequently, they would separate to form two hydrogen atoms, if it was not for the two neutrons. A neutron, i.e., a nucleon with an electron, possesses a seemingly cohesive property in an atomic nucleus. A logical explanation could read that an electron orbiting this atomic nucleus can make the closest neutron's electron to jump over to the adjacent proton as a result of re-coupling. Likewise, an electron placed outside the atomic nucleus could be steered by how an electron in the atomic nucleus moves between two protons that are taking turns to be a neutron.

At this stage, we should be ready to construct a next step before losing our footage. The step could consist of:

A subatomic particle´s handle can be re-coupled

In other words, handles can be decoupled to then latch to other particles. In the above example, we can see how the four nucleons (protons and neutrons) take turns to be neutrons by having one electron in one of the two nucleons (that at that moment is a neutron) jump over and to bind to one of the protons to make it to a neutron. This occurs fast enough so that thc nucleons do not move away from each other. The jumping electron act here as a glue in the atomic nucleus because the protons are competing over it. In other words, we have a stable atomic nucleus.

With this description of these toggling handles, we could boldly claim that electrons surrounding the atomic nucleus do not follow any orbit around it but the simply jump around outside the atomic nucleus as the protons' "string handles" switch between these electrons and the electrons in the neutrons. This interaction between "**inner**" and "**outer**" electrons does not exclude the possibility that these can also change places in larger atoms.

The above would explain the forces that hold together the atomic nucleus and the forces that hold together an atom.

The question is, why do not all hydrogen atoms transform to neutrons through so-called electron capturing. The answer might be that the proton, as everything else, is surrounded by strings, and therefore their attachment toward the electron should already be established in a form of a string chain that prevents the electron from latching directly on the proton. This way, it becomes easier for the electron to latch on this sting chain.
The next question is: How can an electron latch on a string chain already established on the proton, which should be considered as the positron of the nucleon? The answer is: Because the electron has a

greater concentrated mass than the string chain, the electron knocks off the chain, and latches on the end of the part of the string chain that originates from the proton. This is the only end available to form a female-male contact. The subsequent question: "Which of the links the electron can latch on?" can be answered with: "It must be the link that is close enough the proton to form a stable linkage." In other words, before the position becomes stable, quite a few failed linkage attempts can take place. For all the electrons to fit inside larger atoms, some electrons must settle with longer string chains reaching towards the atomic nucleus. Hence, the length of the string chain determines how hard the electron sits, and to which shell it belongs.

Figure 4. Depicts the helium atom as a model how the atom's primary components interact with each other and with the surrounding strings.

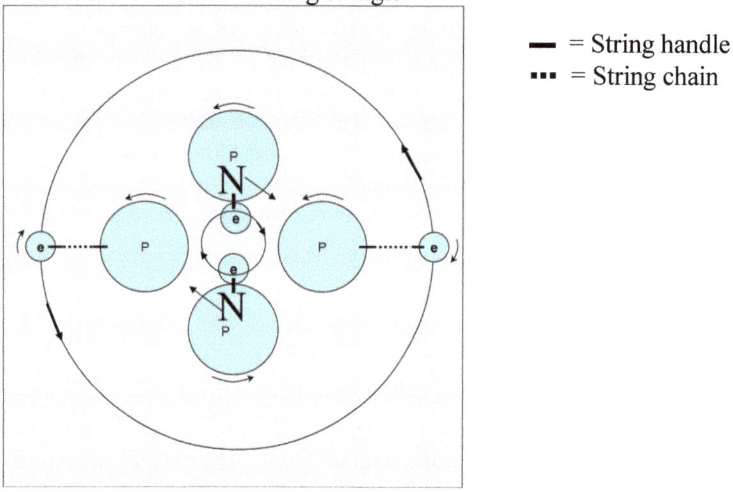

The never-ending movement of the strings hitting against each other jerks constantly the string handles of an atom's outer electrons leading them to temporary become detached from their relatively short string chains to atomic nuclei. The temporarily freed protons in the atomic nucleus create a possibility for the inner atoms to continuously shift between the protons in the atomic nucleus (proton + electron = neutron). This dynamic should be the underlying reason why the alpha particle can exist for shorter moments while 2 handles are constantly available for 2 more outer electrons, and this is what makes helium atom so stable, and explains why 4 hydrogen atoms prefer to form a helium atom and not the other way around. The intra-nuclear handle exchanges should cause the involved particles to spin around their own axel and to sling the outermost electrons from proton to proton around the atomic nucleus. It is unlikely that the orientation on both spin and electron orbits is constant, but is likely a result of the action of latching to surrounding strings (see next pages) and of the kinetic energy/inertness of the subatomic particles. This intra-nuclear dynamic requires also that quarks in the nucleons turn in synchrony with the reversing roll. See details in the text to Figure 12 of Section 4.

Combination of the outer electrons spinning and latching to surrounding string chains should cause waves travelling through the chains hence giving off a recordable signal. Such signals have indeed been recorded.

The above description on how the "inner" electrons jump between protons in an atomic nucleus illustrates also how the property "neutron" moves without a need for the actual nucleon to move.

Section 2 of the book explains why a photon does not exist. Here follows part of the explanation. The higher weight of 4 hydrogen atoms vs. one helium atom is explained by that the energy that stems from fusion of hydrogen is created by emission of photons, and these would stand for the weight difference. But because this book explains the gravitational force as the latching of "external" string chains to electrons located outside the atomic nucleus, as explained on next pages, we can already here see how all puzzle pieces are falling in place. It is obviously not so easy for the external string chains to access the two inner electrons' handles inside the atomic nucleus, and consequently, the weight is perceived lower. And this fusion energy must be transmitted in form of waves through string chains, which is also mentioned in 'Death of the photon', section 2, and in 'Kinetic energy and string chains' in Section 4.

Inter-electron string chains

By definition, electrically charger atoms, ions, do not have equal number of electrons and protons. A key question in this context is why are atoms with eight electrons on their outermost shell so stable that they cannot be ionized, and why other atoms are happy to become ionized, i.e., to borrow or lend out electrons to reach this number of atoms on their outermost shell.

To answer this question, the limits of logic are being tested with following claims. A cube had eight corners. If we think that the eight electrons are placed in these corners, and that the edges are formed by linkage between each corner/electron, which would imply that each electron – except the linkage to a proton – also forms at least three separate linkages to other electrons. In other words, in addition to the already described string handles attached to the proton, each electron should have three other handles that probably fit another string type that is different from the string in the chain between a proton and electron because there should be no inter-electron female-to-male connections. See Figure 5, including the text. Should an electron have a different number of strings of this type, it would consequently entail that the number of electrons on the outermost shell would be different from eight, and that we would have completely different molecules than those we currently have.

And hence, in the inter-atomic relations, it becomes easy for an electron from an atom to stick to form a corner with another atom in the described cube. But such a borrowed electron will never get to form a coupling to a proton in the "host atom". The coupling returns to the atom it belongs, and acts as a stabilizer or the formed molecule. On the other hand, in liquid or gas, other molecules can come in between to cause the lent electron to detach from its proton and create two ionized atoms. Even if the preconditions for an atom to build molecules based on the stable cube exist, any free handle of the electrons will always latch to surrounding string chains (their length can be infinite) that in turn latch to available handles within the same atom or to available handles on electrons on other atoms. We return to this subject under 'Gravitational force' in this Section, and in the text about He_2 (so-called helium II due to its different characters) in Section 3.

A cube is the smallest possible symmetrical form, and the strongest, that can be build up by eight electrons. It is so strong that an atom with an additional proton in its nucleus is forced to have an associated electron on a new shell. The number of electrons and the current atomic shell defines therefore in which geometric shape the electrons form connections to each other and subsequently also the electron's so-called energy level ('s', 'p', 'd', etc.). Moreover, an electron should not be able to spin around its own axis if all these string handles are occupied within the same atom. Electrons with "available handles" are naturally attached looser, and they can also form linkage to surrounding string chains, and consequently also to other atoms.

It might be of interest to know in this context that a radon atom is heavier than a lead atom. But because of the radon atom's outer shell is built by 8 electrons, it cannot latch to another radon atom and remains in gaseous form. Radon tends to decay by emitting alpha particles in so-called radioactive radiation. This demonstrates that there is a limit for how large an atomic nucleus can become before some of the many alpha particles of the atomic nucleus, at a regular frequency, happen to be left outside; both in exchange of the inner electrons and in forming linkages to the outer electrons. Read also about the so-called "tunnel effect" described under 'Possible explanations to phenomenon within quantum physics'.

We can also note that the two electrons in an atom's first shell are seated securely. Because of the stable union between these electrons and related protons and neutrons, which is already explained in Figure 4. The hydrogen atom is – for the same reason – uppermost unwilling to borrow an electron to secure two electrons on its electron shell. The same figure depicts also clearly why an atomic nucleus with more than one proton must have neutron to hold together. In the light of the above arguments about inter-electron string chains, the two "outermost" electrons in Figure 4 can exist on the same side of the atomic nucleus. Especially so, if they have a mutual coupling at least to one string chain, and/or if they are linked to another helium atom's two electrons, hence forming He_2. See Figure 10 in Section 3.

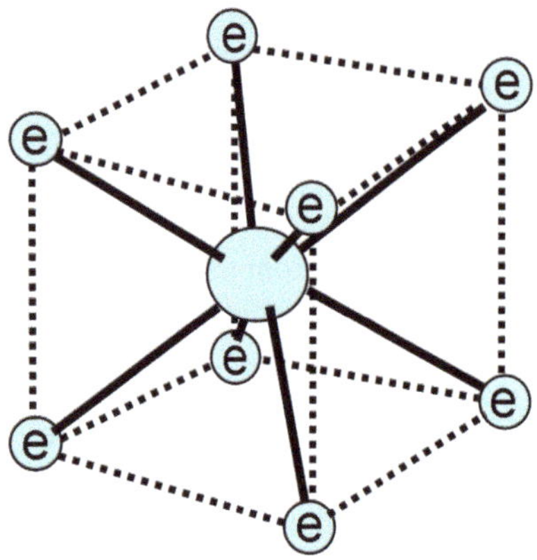

Figure 5 shows an atom, which has 8 electrons in its outermost electron shell, and how these are connected to each other by the means of string chains (dashed lines), which are formed by strings from the dark matter.

An atom's outermost electron shell can also consist of a number other than 8 electrons. But the fact that an atom, which has an outermost shell of 8 electrons, cannot bond with other atoms tells us that each electron has 3 string handles, no more no less.

With these 3 string handles, 4 electrons could also form the corners of a tetrahedron (pyramid) in an outermost shell. But such an atom apparently becomes more exposed to the environment, probably because the attachments of the string chains to the electron become too weak at smaller angles between the string chains, possibly due to the handle's location on the electron. The cubic form is the smallest possible symmetrical form and thus the strongest, which can be formed by eight electrons. With a number other than 8 electrons in the outermost shell, the atom, as understood from the above, gets free string handles towards the surroundings, and can then form molecules.

The electrons inside the depicted shell are not in the picture, but those of their string handles, which are not coupled internally, couple to the electrons of surrounding atoms. This is what makes up both gravity and the Van der Waals force.

The ability of strings to latch to each other

A question that could have been asked earlier in the book is why are the strings interested at all to latch to each other. However, this is a question with even a greater relevance once the book's theories about subatomic string handles are reviewed. The question can be answered by thinking of the strings as uneven objects, and because of congestion, the strings constantly collide with each other. Their shape determines whether they can build subatomic particles, as discussed above, or only latch to each other in chains that eventually latch on to the already formed subatomic particles.

Based on the above, theories of the so-called **Higgs particle** and **Higgs field** appear like floundering in darkness. However, a sense of respect remains for all the work behind the theories with their origins in 1934 when Yukawa presented his theories on nuclear forces. Based on the concept presented in this book that subatomic particles consist of strings, it is not possible to claim that one string or a random aggregate of strings would be more essential to the particle's mass than other strings or aggregates. What makes a subatomic particle "accessible" to become part of an atom and thus reach a recordable mass, are its handles.

The finding of Higgs particles could therefore be a subatomic particle's string handle that is attached or detached leaving a mark in form of a wave. The definition of Higgs field fits as the definition for interaction between string handles and string chains, or between the unevenness/gluons that can make for example quarks to stick together.

Summary of string handles

1. Because the electron and proton (a nucleon) are linked together over a – in this context – long distance, they must have something that holds them together. And because of the different distances found within an atom, there is a reason to assume that this linkage consists of a string chain because these can be of different lengths. It is also reasonable to assume that the links, the strings in this chain, have contact-forming "male" and "female" ends, or alternatively, male and female strings in the chain. The reasoning behind this is that a proton's and electron's contact-forming ends are not identical, which is further based on the notion that an electron can only latch directly to a nucleon and form a neutron, if the quarks in the nucleon are facing the other way. See also Figure 12 in Section 4.

2. The electrons have also other types of handles. These can be coupled together. Due to the variable distance, which must be assumed to exists between electrons, also the inter-electron couplings must hence consist of string chains. Because the electrons are identical subatomic particles, the ends of such string chain must be identical. This type of string chain does therefore not have a specific male or female contact end. On the other hand, every other string in such chain could have male contacts in both ends, and the other could have female contacts in both ends. An electron can form a connection to up to three other electrons which is shown by the inert gases with their eight "fixed" electrons.

3. The structure of the noble gas atom also explains how the electrons connect to each other when atoms in other elements form molecules. A molecule is normally completed when there is no longer a vacant string handle for any of the electrons in one of the outer shells of the atoms. I.e. when this atom has gained eight electrons. The fact that two hydrogen atoms are stuck together also depends here on the fact that each of their electrons has connected to each other. But each of their electrons has two more vacant handles. When the two hydrogen atoms bind to an oxygen atom to form a water molecule, it is one of the two remaining handles on each of the hydrogen atoms' electrons that connects to the outermost electron shell of the oxygen atom, which then gains a total of eight electrons. The description above also explains why the water molecule has both hydrogen atoms on the same side of the oxygen atom.

4. The fact that an atom's electron does not use more than one of its three handles when connecting to other atoms is probably based on the geometry, which would lead to too great an angle on an extra connected string chain, and thus become an unsustainable connection. The free handles therefore connect via string chains to the surroundings and thus explain gravity. See "Gravitation" in this Section.

5. At the absolute zero point of the temperature, the atoms do not move, and an otherwise geometrically untenable "double-coupling" between electrons can then hold. This leads, for helium, to the fact that no handles are free on the electrons, and that gravity thus ceases. See "The Evidence: Helium II" in Section 3.

6. A neutrino, not yet mentioned in this book, does not get caught, and hence it behaves like a particle without string handles. It appears when a proton is fragmented to free the nucleon's three quarks. See "The neutrino, does it exist?" in Section 4.

7. It might be an appropriate moment to mention that repulsive force between subatomic particles, such as protons, is likely to be nothing more that the collisions failing to latch particles together because their handles do not fit together.

Gravity

As described above, chains get hooked on strings in different parts of an atom to subsequently create the different forces that build and control the atoms' world. It is quite easy to understand how strings, or chains made of strings, latch on various elements of an atom because the subatomic particles are made up of different strings or their combinations.

The preceding summary explained why an electron is likely to have three string handles that form connections through string chains by linking to other electrons' corresponding string handles. The fourth string handle of an electron is different, and forms a linkage to a proton through a string chain consisting of a different string type. Since each atom has more string handles of the former type than what is used between the electrons in the said atom, can "external" string chains latch on these, form couplings to electrons of other atoms and eventually form molecules, as described above. The latter is probably only possible between string handles in the outermost electron shell. It is also possible that this coupling can be conveyed between electrons in inner shells by indefinitely long chains of strings, independent of the distance. See Figure 6a. In the following, and based on the above, the gravitational force will be explained.

By observing our planet Earth as a great example of how gravity works, it becomes clear that when a body, such as planet Earth, is filled with these string chains, they must have a radial orientation. This is caused by "congestion" and because the string chains are linked to the string chains located outside the planet.

Object on, in and outside the planet become therefore affected in the direction of the said string chains. **The never-ending movement of the strings hitting against string chains causes (gravitational) waves along them pulling the electrons' string handles with a thrust possibly amplified by the "spin" described in Figure 4.** The gravitational force is "born". It is therefore easy to imagine how atoms are being pulled by the string chains towards the center of the Earth. Any movements inside the gravitational field disentangles, with high intensity, each affected electron from an external string chain, to then latch on to another. A planet with larger mass than the Earth lends higher density of string chains and subsequently a higher thrust. Hence, the thrust against an atom is limited by the number of available string handles in the atom.

Continuing with planet Earth as our example, its rotation around its axle is likely to cause a weak bend of the radially oriented string chains. Consequently, an object should not be pulled towards the Earth in a completely radial line. See Figure 6b. And what this book **defines as gravitational waves** should have a higher speed than light because light waves, as described in Section 2, lose time due to distances between individual strings and string chains.

A relevant question would be why would not a linkage to a string chain pull an object away from the Earth. This can be illustrated by a person holding with one hand a short rope attached to the closest wall, and with the other a longer rope attached to a house located further away. Movements in the short rope create larger thrust than movements in the longer rope, if both ropes are weightless. Because the longer rope nevertheless has a certain amount of thrust explains why the moon exerts its gravitation on us, as illustrated by tidal waters.

Gravitation is experienced as a constant force due to the constantly occurring re-coupling of string handles to string chains, even in movement in direction towards the "center of gravity". But since the re-coupling speed of the string handles has its limitation, gravitational force will not be applied on objects that are moving at a certain speed towards the center of gravity.
To conclude about gravitational force: it binds also atoms/molecules to each other. And because there is no limitation to the length of string chains, gravitational force can cover infinite distances.

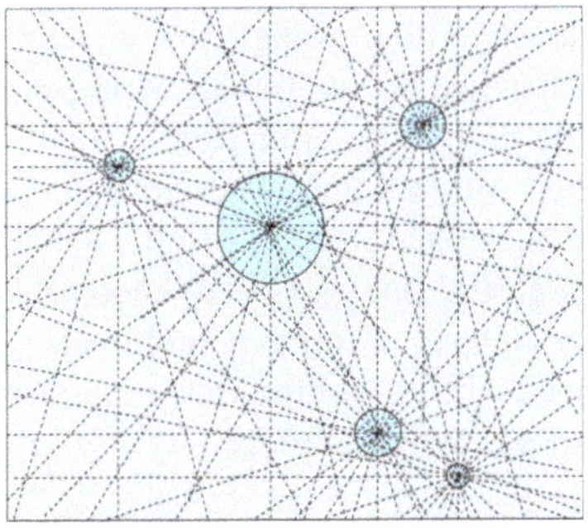

Figure 6a. Shows string chains that cover distances between planets and stars.

Figure 6b (bottom). Shows how gravity/string chains bend around a rotating body, such as planet Earth, and how the gravitational field varies as a response for example to the moon. Tidal water illustrates this clearly. '1', '2' and '3' represent G forces of different sizes, in this case, on the Earth. The side of the moon with the largest mass stays naturally facing the Earth. The gravitational force of the Earth affects partially the moon's own gravity. This is illustrated with 'a' through 'd'. If the moon's mass was equally divided, the minimum force of gravity on the moon would be 'a' (because of the Earth's rotation), increase to 'b', 'c', to be at its maximum at 'd'.

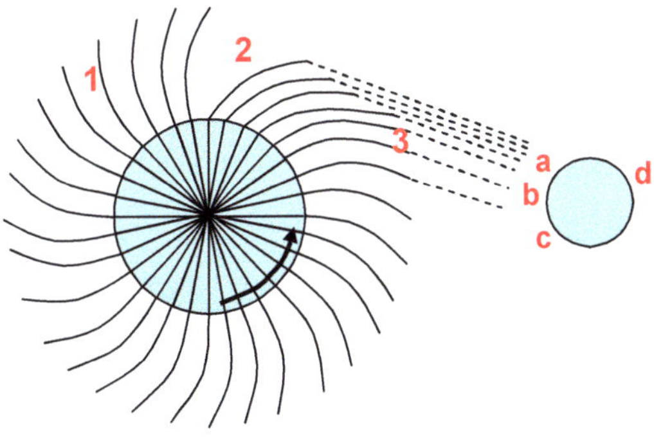

Nothing is new under the sun, and the history shows that many scientists, including **Airy**, **Stokes**, and **Planck** at the end of the 1800s, thought that "ether" (strings and string chains of this book) was dragged around the Earth, as illustrated, but with respect to the Earth's movement at 30 km/sec around the sun. This was used to explain discrepancies in angles for stars (stellar aberration) compared every 6 months. Light was considered to be waves through the ether, which coincides with this book's conclusion. See also 'The constant speed of light, and Einstein' in Section 2.

Magnetism

Electron shell of an iron atom causes these atoms form small units, crystals, allowing a few electrons inside them to move freely. When observing a magnetized piece of iron, an assumption can be made that each such unit has established an excess of electrons at one end. Because the electrons, as described above, are linked to string chains that connect them to protons, then there should a bouquet of strings that have lost their contact to their respective proton and are beaming out of these electrons.

At the same time, there is an equal number of protons that have lost their coupling to their string chain and subsequently their contact to their respective electrons. When the distance between these strings gets shorter to the string chains of the electrons in the adjacent magnetic unit, these units form "circuits", instead of connections being formed between the ends (poles) within a unit. Therefore, it could be assumed that images of a magnetic field show string chains that radiate outwards from each pole on a magnet, sting chains that are directed toward each other through the sides, and will become linked. By first establishing single smaller magnetic fields, these chains of strings can explain how a piece of iron can become magnetized by the Earth's weak magnetic field. The rest of the iron becomes magnetized through the associated string chains and circuits of these small magnetic units.

Because the described bouquets of string chains originating from the two poles connect to each other, the interlinking chains should have both male and female connections. So, when a string chain originating from a pole meets a string chain from the other pole, they are linked together.

The magnetic force ends, like the gravitational force, both by the noise caused by strings hitting against string chains, and by the thrust caused by the constant re-coupling. The "spin" described in Figure 4 is also affecting these electrons which could also contribute to the thrust.

"Magnetic" repulsion does not exist

When two identical magnetic poles are pressed toward each other, what we call for repulsion becomes obvious. This term can be deceptive and misleading. Apostrophes in the headline indicate that repulsion cannot be based on inverted magnetic pull but it occurs because of congestion among strings. This can be described by attempting to place an object in a place already occupied by something, which makes the strings palpable.

On the other hand, if one of the magnets is turned around, the string chains from one magnet's electrons will connect to the protons in the other magnet, and the other way around. Both magnets act now like one magnet, and in this case, the above-mentioned congestion among strings is absent because string chains become linked and pass through both magnets.

String chains that form the magnetic field can also pass through nonmagnetizable substances, such as air or organic material. And if the magnetic pull becomes large enough, its string chains become noticeable also to the nonmagnetizable object. In experiments using strong magnetic fields, the first completed as early as in 1939 by Braunbeck, and by other at the end of 1990s where, among other things, a small frog was kept levitating. Such strong magnetic field can be illustrated as hammocks made of string chains which possibly partially acts as a hinder for string chains mediated by the gravitational force.

According to this description of magnetism, a material can be magnetized if it is able to retain the electrons that have left their protons. Static electrical exposes an object to same type of "magnetic forces" as long as the affected object is kept electrically isolated.

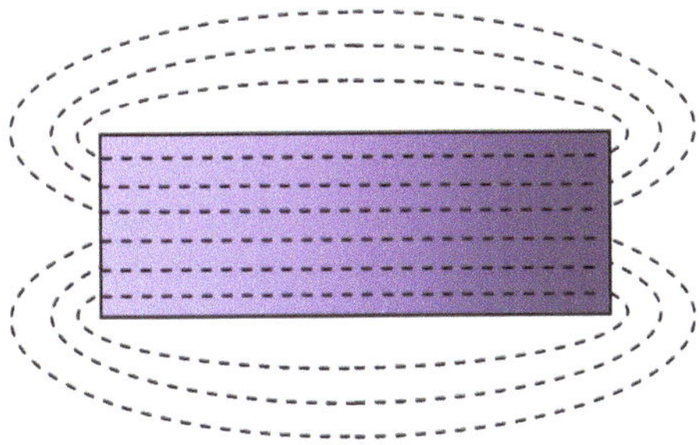

Figure 6c. A magnet with a magnetic field of string chains that could be extended positrons, as defined in Figure 12 with the description in Section 5.

Electromagnetism and electricity

With the description provided for magnetism, it becomes easy to understand how a magnetic field with related uneven distribution of electrons creates what we refer to potential differences in an electrically conductive material in this field. When these are evened out, it causes a migration of these electrons, a so-called electric current. It is something a generator generates. The fact that electric current in turn generates a magnetic field should logically thinking be the opposite, i.e., that migration of electrons pulls the string chains attached to them in one direction at the same time as the string chains originating from the abandoned protons will turn towards the incoming electrons.

A transformer utilizes a combination of the above, with incoming current in a so-called primary coil surrounding a "body of iron" that we can call iron core. A secondary coil for outgoing current sits outside the iron core. When current is led through the primary coil, and the iron core becomes magnetized, any "lose" electrons of the iron core will flow in one direction creating a current in the secondary coil. Because the distance to move inside the iron body is limited for the electrons, there will be no more current in the secondary coil once the electrons inside the iron core have reached as far as they can go. Therefore, the current in the primary coil must be constantly altered, such as in alternating current, for the secondary coil to continue to produce current. The following is possibly a new way to see the current generated by the secondary coil: One round in the secondary coils senses the distance that the electrons have travelled inside the iron core, and will affect the pressure of electrons of the round. Two rounds create twice the pressure, i.e., double the current. This can be compared to harnessing horses in a series. Concurrent decrease in the total ampere by half is caused by the doubled (electrical) resistance which is proportional to the length of the wire.

An electric potential is created by the following different factors: **1.** A magnetic field, as described above, is catching and taking along electrons. **2.** In static electricity, one material has higher attractional force on electrons than another adjacent material. It is easy to see that the said attractional force can be different if one considers that atoms' and molecules' appearance can vary greatly with subsequent differences in how securely an electron is attached. **3.** In chemistry, differences in the attractional force are also driving factors. **4.** Lastly, there is the so-called photoelectric effect where light (or other high-energy waves passing through string chains) knock off electrons from their

attachments in an atom. If these lose electrons can be absorbed by a material where they do not come off by the same impact, a potential difference is created.
An established potential difference aims to even out the difference but naturally not in the same location where electrons are getting separated. This is the driving force behind the electron movement. How do they move, from atom to atom inside an electric conduit? It is essential for this to occur that an atom that has lost at least one electron will meet another atom. But why would an electron in another atom let go off its existing couplings to then form a linkage to the atom that has lost an electron? The turbulent circumstances that exist at that level, with or without potential differences, answers this question. The turbulence is "one-way" in this case, and a new electron coming from the opposite direction occupies the empty spot before the original electron has time to come back. It is easy to see how affected atoms are forced to move which generates both heat and increased resistance for the current being conducted. Furthermore, the current creates electromagnetism which also causes increased resistance at electron transfer.

Because each electron is busy forming their connections to the proton and the closest electrons before is sits "securely", there is time for the next electron to enter before this occurs. Therefore, it is easy to see that it is not the securely attached electrons that are forced to move aside. It has also been verified experimentally that electrons follow in succession in an electric current.

As the name suggests, superconductivity does not have much of resistance. Observations of these materials that often needs to be cooled down to almost 0 degrees Kelvin have revealed that their electrons form pairs, so-called Cooper pairs. This book argues that such pairs are electrons bound together by more than one string chain. Therefore, the conclusion could be that electrons in transition "surf" on these extra inter-electron couplings. As current is conducted a superconductor should become filled with electrons, one for each chain. This explanation makes it easy to see that the effect on atoms in this material is minimal. Most importantly, the strenuous on and off couplings to protons are not needed. The electromagnetic resistance and heat generation are consequently minimal. Superconductors are also described in Section 3.

Otron plate; String handles and intersecting magnetic field

The target of my journey was an otron plate. While on the road, have analyzed the reality, and made many conclusion. The idea was that these would help me to develop the next-generation Proton Plate. But now the journey itself has become the target, and the following clarity has emerged:

If an electron is removed from a neutron, a proton is formed, possibly because of a positron "hurrying" there. And if we remove this positron, we get – if not an "otron" (my own term used in this book) – then three quarks. The question is whether they can fulfill the dream of the Otron Plate. Other particles that appear as completely closed string bolls, such as neutrinos, without handles could be used, if this is more than just a wave. How would these or the quarks of the "otron plate" then be captured, kept and stored so that they, in a form of a shield, would break the string chains that are reaching toward the Earth. However, the most important contribution of the analysis in this section is the conclusion about what creates the force we call gravitation. Once again, this time after a carefully completed analysis, it looks that I was, after all, on the right path in late winter 1974. A collection of subatomic particles not including the electrons, should be able to carry out the task, i.e., break the string chains reaching toward the Earth that mediate the gravitational force. The outermost electrons' cube configuration in inert gases is of no help in this context because external string chains can latch to any available handles on electrons located inside.

An exception among inert gases could be He_2 (see Figure 10 in Section 3) because this molecule has no available string handles when two double string chains are in use. It would have been an option had it been stable and had it been possible to keep in on one place in a magnetic field, possibly by freezing it down. Another alternative is to try building a molecule with a stable form and that has no available handle either.

Also, it might be possible to use magnetism to "cut off" those string chains that mediate the force we call the gravitational force. One or more, possibly intercepting total magnetic fields would hopefully make it too crowded for the string chains that mediate the gravitational force. The section 'Magnetic" repulsion does not exist' explains that the magnetic repulsion is caused by congestion among strings between the poles. At the time this section was written, in earlier editions of the book, the above experiments, one of them with the frog, were not known to the author. These experiments prove the existence of string chains that can act like "hammocks" when present in sufficient numbers. The question is whether they can reach sufficient density to cut off the gravity-mediating string chains.

Conclusion of this section

One should remember that the above statements are theories that can neither be proven nor disproven. However, the logic behind the theories is based on the following items that most of should be able to agree on:

1. All different subatomic particles must have been made of something. We can call this something for strings. A concept of '**vacuum state fluctuation**' has already been established. It fits well into the concept of this book.

2. Is seems highly unlikely that the string composition forming a subatomic particle would be so perfect that there would be no interaction with other strings.

It is noteworthy that is looks like the four natural elements, gravity (including Van der Waals force), magnetism, the strong nuclear force (interaction inside an atomic nucleus), and the weak nuclear force (interaction inside a neutron), are explained in this section.

Allegedly (just like in the religion) it is not possible to use logic, such as above, at that level. And in those cases where shortcomings have been supplemented with new theories there has been a risk to end up with an inconsistent and less reliable patchwork. It is better to retrieve and try a "new entry into the labyrinth".

Research on atoms has revealed many fine theories about subatomic particles, such as pions as the inter-nucleon particles with cohesive properties inside an atomic nucleus, and about gluons as the inter-quark particles with cohesive properties inside a nucleon. Gluons, on the contrary, are thought to exert an effect between quarks of the adjacent nucleons. Careful studies and analysis of the results from the experiments have provided explanations that include even more particles, such as bosons, leptons, fermions and gravitons. Gravitons are thought to explain gravity but the particle has never been found. Formulas and laws, such as Coulomb's law have been produced to describe in detail the repelling or attracting force between two electric charged particles, provided that they do not move in relation to each other.

This book is not directly opposing any of this, except for explanation to gravity.
Quarks could be small "string bolls" that stick to each other either directly or with a help of gluons that could be short string chains. Up quarks and down quarks can be attributed to how they become linked to either a positron or an electron. See also Figure 12 in Section 4. Why an electron does not play a greater sole in the theories explaining the co-existence between protons and neutrons could be questioned. But if we boil it all down to the string level, it is here where everything begins and ends. At that level, we are left with concepts, such as gluons and electronic charges, far behind us. **Because, what it is *all* about, and what the four natural elements all boil down to is this:**

Moving strings that latch to, or do not latch to, each other. (Attraction or repulsion.)

Strings, as stated, include the strings that build up the subatomic particles, and subsequently also the strings with ends that are described as string handles in this book. The concept "strings" include also the strings/string handles that surround and possibly form connections to string handles on particles. Everything is strings. It is simple. It is also logical to think that the deeper under the surface of something complex one looks, the simpler things get.

John Archibald Wheeler was "on the edge of" understanding this simple principle of string level simplicity already in 1955 when he described something he called for "**quantum foam**". This has also been calculated mathematically using calculations that seems unnecessary complicated because they observe theories about time and room.

The history of particle physics is full of theories of various "body parts of an elephant", and nobody understand what they see. Yet, it is mighty **impressive that so many islets of knowledge have been drawn through the years** despite the lack of a coherent picture depicted in this book. At the same time, it is rather odd that nobody has succeeded to "paint" the whole picture. Lorentz's and Einstein's assumptions on photons and time can be blamed for this since they keep the door closed to the solution for this puzzle. The messenger, a full-size elephant in a form of helium II, has been knocking on the door but no one dares to open. Einstein's theories will be discussed in the next section and helium II in Section 3. The **following summary** is important for understanding the content of this book. **On the following pages, the spearhead will bust "Einstein's world".**

Role and characteristics of the strings

1. Building blocks for all matter are composed of strings. Hence, the mass is derived from all included building blocks, and these interact with the surrounding strings. (Get latched or get not latched.)
2. Despite their mass, the strings outside matter are **not affected by the gravitational force** but they create and mediate gravity between matter – *like a chain shaking between two knots.*
3. The gravitational force does not increase the density for strings that are outside the matter.
4. On the other hand, string density increases towards a center of presence of matter, such as a planet or a stone. This can be compared to spokes in a bicycle wheel. In a black hole, this relationship is set out of order upon annihilation or destruction by pressure of many molecules resulting in fewer handles reaching toward strings.

The above items show why matter is bound to surrounding dark matter. This explains many phenomena, which are dealt with later in the book. Next part, Section 2, takes us little further out in our chase for zero gravity by scarifying the existence of a photon.

Section 2

The Photon Does Not Exist

and neither does the phlogiston*.

-It is easier to believe in a lie,
heard one thousand times,
than in a piece of truth, heard for the first time.

Unknown

*) An imaginary substance that the bannermen of science believed in for centuries, despite evidence of cracks in the logic.
Today, we are facing the same issue with the photon. Should a photon exist, then interpretation of the experiment (carried out by scientists, and referred to in this section of the book) would state that a photon, completely illogically, would know what another photon is doing.

The constant speed of light, and Einstein

Here we will provide a historical summary of what has been shaping our understanding of the speed of light and the so-called photons. An important experiment was run in 1887 based on the concept "ether" (dark matter). Despite the wide-spread and long-lasting suspicion on the concept dark matter has the modern science been based on this experiment. It is a paradox that seems to have passed unnoticed. And that the science in this field has ended up in a corner with no way out is unfortunately understood as a natural course of matters. A life-saver would have been the knowledge about the role and characteristics of the strings, as presented at the end of Section 1.

Background

This book claims that the photon does not exist, and that all electromagnetic radiation – light included – is a wave motion passing through strings. Already in 1800s, it was understood that light was waves traveling through strings that was then called the "ether". **Jean Fresnel** suggested in 1818 a theory that this ether is not only surrounding us but it also penetrates all matter. It is completely in line with this book's theory about strings and string chains. In 1864, **Maxwell** calculated the how fast light travels through the ether. He suggested that light consists of waves that travel through the same medium which is [sic: the cause] of the electric and magnetic fields, and that all three are electromagnetic in their character. **BUT** the assumption of ether as the medium for light was unexpectedly shaken by an experiment that concluded that light is sent out at equal speed on both directions, forwards and backwards, in relation to the Earth's movement at 30 km/sec. around the sun. The experiment by **Michelson and Morley in 1887** was conducted by arranging a set of mirrors to reveal any possible interference of light waves if the light sent out from one source would take a longer to travel in the direction of the Earth's rotation. No difference was observed, and this is the data which caused a lot of headache among those carrying the torch of science of the time.

This book defends the explanation that ether is attached to adjacent matter – also mentioned in **Stokes**' **and Planck**'s theory from 1899 with an assumption that the density of ether increases with increasing gravity. **Lorentz** was another scientist working with the same problem who unfortunately came to the conclusion that the density of ether would then become too high to explain a relatively unchanged speed of light. **Because the role and characteristics of the ether/strings,** as summarized at the end of Section 1, **were unknown, the development took a wrong turn.** (And leading the turn is today's elite within the field trying to find a way out with new theories.) Also, the results obtained from **Michelson's, Gale's and Pearson's** experiment in 1925 were considered to contradict Stokes and Planck. Lorentz published his explanation in 1904. He had rejected Michelson's idea, being certain that ether is completely separated from matter, and thought that the experiment's physical frame must be something that moved at a speed of 30 km/sec. into a new position in the ether. And that this was the reason why it no longer was possible to count on having the normal laws of physics to apply to the instrumentation that was following the waves through ether.

With his transformation equations – a further development of the Galilei transformations created by Newton to measure time and space on two places that move in relation to each other – he suggested that the problem was solved. The equations resulted in a concept called **length contraction.**

According to this concept, the faster something that can be measured in the direction of motion travels, the shorter it gets. The concept **time transformation**, meaning that time passes slower for an object the higher it's speed is, is another side of the same thing. Lorentz was not alone expressing these ideas but his name got assigned to these ideas. Read more about '**Lorentz factor**' in the Annex at the end of this book. This was the back drop in 1905 when the destiny of the world's science was decided on an open battle field. It was not decided by a comprehensive vision in the lack of essential knowledge. Not even the existence of an atom had been established yet. 'Kinetic energy and string chains' in Section 4 is of interest in this context. It demonstrates how ether is both anchored in and follows along the adjacent matter. See also Figure 6b in Section 1.

Einstein's Special Theory of Relativity

The father of "the coup" was, in this context, a relatively unknown young man with the name **Einstein.** He had been solving puzzles with Maxwell's, Lorentz's and Planck's theories and equations, and built them further. Einstein was only 26 years old, and the researchers who had struggled with the problem for many years were exhausted for idéas. That someone "impartial", like Einstein, took over and presented a solution, with ingredients from most people involved, could have been tempting for the researchers to reluctantly accept. The problem was that Einstein based his solution on postulates that he made up for solving the problem. Using these postulates he built mathematical models, which resulted in rather populist conclusions. The researchers protested, but Einstein, who was younger, survived them. **See "Einstein's two postulates"** in the Annex.

With these postulates he made the spreading of light independent of ether as medium, which in turn meant that the light must consist of particles, on which Planck's concept; **light quantum, later called "photons"**, came to be used. Einstein claimed in his famous publication from 1905, **'Special Theory of Relativity'** with his so-called field equations, for instance that the relative speed of an object in relation to the light is constant, regardless the object's own speed. **Observe that an incorrect assumption, for example on time and space, still is incorrect, even if the assumption can be described mathematically.** Despite that; This field of science eventually became a victim of the worlds greatest brainwash, that has overruled both logic and objections, a heavy fact that is easy to verify. The need for a new generation of scientists, that dare to stand up for logic, is therefore urgent.

As is; it is not surprising that logical problems, as the following emerges: If someone was travelling at a high speed, for example at half of light's speed, then one could speculate that the light would still pass this traveller and leave him at the speed of light. Because this logical dilemma cannot be patched with just any patches, Lorentz's time transformation, "Emperor's new clothes", provides a customized solution that fascinates people who have cautiously applauded to it for a century. **To align the constant relative speed of light with mathematics, it was necessary to adopt the claim that the faster a traveller travels, the slower the time passes for the traveller.** It was believed that Einstein was able to prove this with his famous thought experiment about light in a train. See Figure E.

But if the atoms of all GPS satellites would show a consequent slowing up of the displayed time, this could be explained by a more intensive pressure/flow through of string chains, which must have an impact on subatomic processes for both people and instrument measuring time, and this without slowing down the time.

Relevant questions to anyone who thinks he or she has seen the Emperor's new clothes: Do we have, based on different speeds and direction of motions for objects, different time systems in the universe? And what should the speed of an object be measured in relation to?

Figure E. Einstein's thought experiment that would show how time travels slower for an object on move. The following images illustrate Lorentz's theories copied by Einstein.

A. An observer inside a train car sees light beaming down, become reflected in a mirror to then go straight up.

B. Another observer standing by the passing train sees the light to travel a longer distance during the same period.

Shows the same part of the train carriage on three different occasions when the carriageis moving to the right.

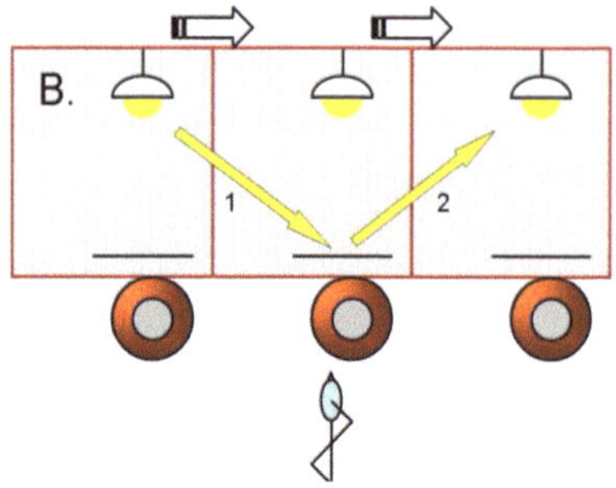

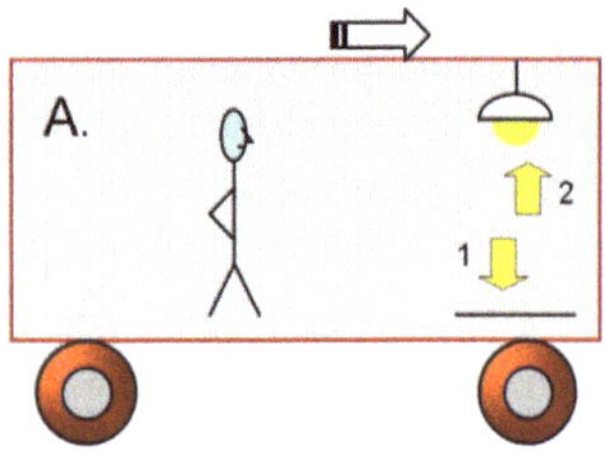

Because Einstein's assumed that the speed of light is constant then the time must have slowed down for both observers and for the light inside the train car, than for the observer outside the passing train. However, this book claims that light is made up of wave motions at a constant speed through dark matter, which in this case is bound to the train. The resulting speed of light, is therefore affected by the speed of the train.

Regarding the above thought experiment.

All objects' movements are relative to their surroundings. The fact that the path of the light in the experiment is relative to the surroundings of the train carriage means that time passes more slowly in the train carriage also when it is stationary in relation to, for example, an observer in a passing train carriage. How do Einstein's lackeys explain it? And since the driver of the train does not live longer than others, despite the long horizontal displacements of the vertical springing movements during all the driver's years of service, we can conclude that the speed of time is not variable, as this experiment was otherwise intended to show.
Concerning the propagation of light relative to the dark matter (strings); see Figure EX1 This book argues that light consists of waves traveling through the strings/string chains that we are always surrounded by. The book contains many supporting examples. Also in this case, we can build a theory that can be based on the above. We have two solutions.

Figure EX1. Solution 1. Solution 2.

Shows the same part of the train carriage on three different occasions when the carriage is moving to the right.

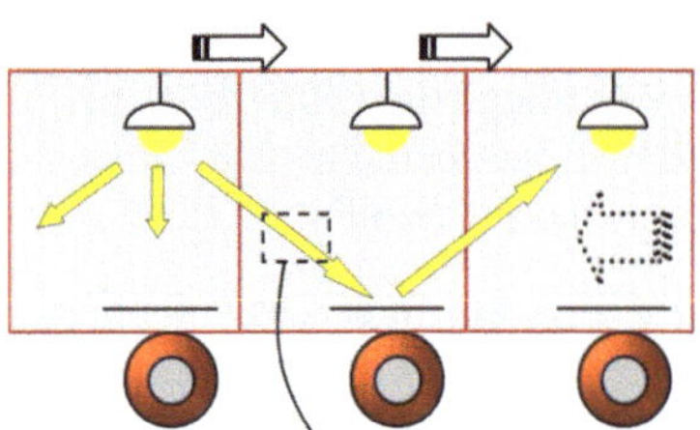

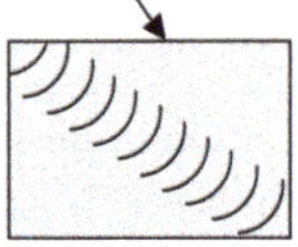

1. A flow-through of strings. In this case, the light gets a longer distance. (Like swimming upstream.)

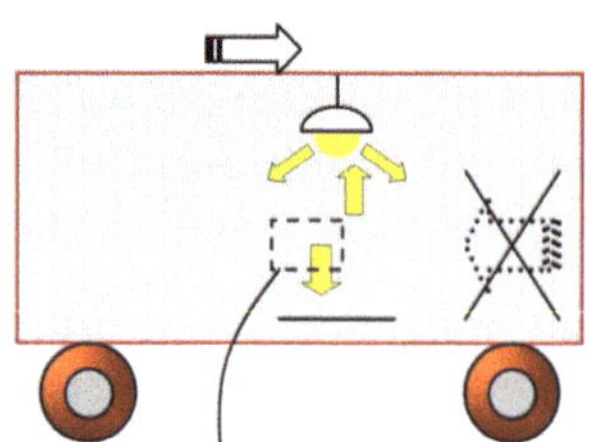

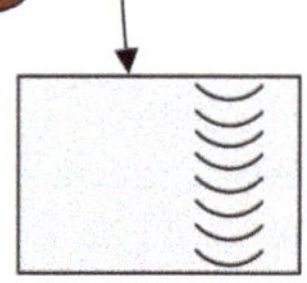

The dashed broad arrows symbolize oncoming dark matter. Crossed arrow represents that we do not meet it, but that it follows along.

2. Strings are bound to the surrounding matter.

In Solution 1, the train car is moving *through* an ocean of strings. The speed of light is constant through the "ether" according to Maxwell but because light *meets* the strings, its medium, it must travel a longer distance. It must take longer, and can be compared to swimming upstream. Light travels at a constant speed through the ether. (The speed of light is *not* always constant but it is dependent on the medium it travels through.) A laser that travels through gas at near the absolute freezing point demonstrates this with a clear retardation.
See **https://www.youtube.com/watch?v=EK6HxdUQm5s**)

If Solution 1 applies, i.e., string chains are not attached to adjacent matter, we get the following: Because the speed of light is constant, and the light source inside the train car already has a speed, the *vertical* speed against the mirror be slower than it would be otherwise. And if technically possible, both observers, one inside the train car and the other standing still outside, would be able to observe this without any slowing down of the time.

If Solution 2 applies, then matter (an object) captures along adjacent strings, which is the most logical option. This will be re-visited in Section 4, 'Kinetic energy and string chains'. This is why the vertical 'beams of light' can hit the mirror. This can be compared here to diving or swimming in a swimming pool on a moving boat.

Both solutions are less far-fetched than the theory of slowing down the time, which indeed is a rather unscientific concoction. Something to think about: An observer inside a train car could at any time reach his hand out to hit a hand of the observer who watches the train to pass. Both would feel the pain at the same moment, and therefore evidently, they live in a synchronized world. However, the risk is real that someone would claim that between two such handshakes, time passes slower for the observer inside the train car. But what about the wheels on the train? Does for example the side of a wheel that at any moment is in contact with the rail exist in a normal time, while the rest exists in another time?

See Figure EX2. Such like Einstein's though experiment, this though experiment is based on the principle of constant speed of light and a train car (that moves to the right), which should delight any conservative puritans. It seems evident that strings (dark matter) in the surrounding get dragged along, as described in Solution 2 for Figure EX1, and the example in Figure EX2 is just a theoretical exercise on this solution.

Figure EX2.

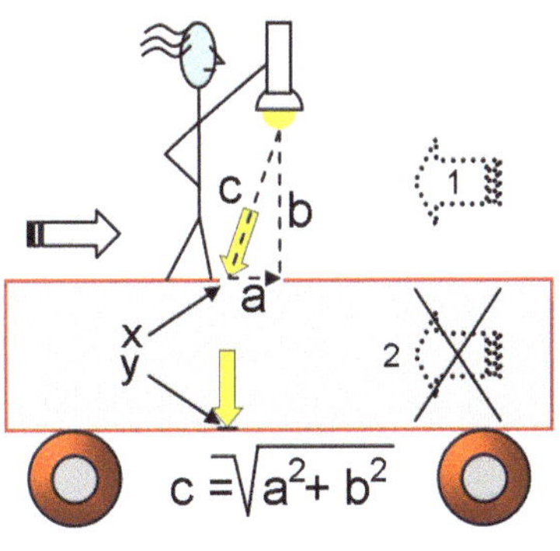

a = The distance the train travels during the time it takes for the light to reach the ceiling of the train car.
b = The distance the light travels while the trains stands still.
c = The distance the light travels while the trains moves.
The circumstances are explained with Pythagoras' theorem

A measurement of the location of "x", should show that the faster the train goes, the further back this location moves, whether the light is photons or wave motions through dark matter. **If the measured location of "y" is no further back than the location of "x", then this proves that the light is wave motions through dark matter, and that this accompany adjacent matter.**

1. A flow-through of dark matter. 2. Dark matter is bound to the surrounding matter.

See Figure EX3 and the associated explanation of variables. A beam of light that is sent out from an airplane for example, leaves the airplane at the speed of light. The problem is the claim that the *relative speed* of light in relation to the airplane is *also* the speed of light, in other words, 'b' is similar to 'c'. To get the argument to align with the mathematics, it is said that the faster an object travels the slower the time passes. It is both contradictory and terrible to assign a speed a variable time unit, and then argue that this speed would be constant. If we disregard the above and use the time factor 't' in the equation; t(a + b) = c, then we claim that time goes slower *also for the light* that leave the airplane! To prolong seconds for a photon with the help of the time factor is for no help because the relative speed of the airplane in relation to the light is assumed be the same also for light that has not been sent out from the airplane. The alternative; t(a) + b does not help either but results in a speed that is faster than the speed of light. The remaining solution has t = 1 and b < c. In other words, the reality is not unreal. We all move along one and the same time line.

Figure EX3.
a = the speed of the airplane, b = the (Note!) *relative* speed of the light that is sent out from the airplane,
c = the speed of light, t = time factor (e.g., = 0.5 applies when light travels "half speed").

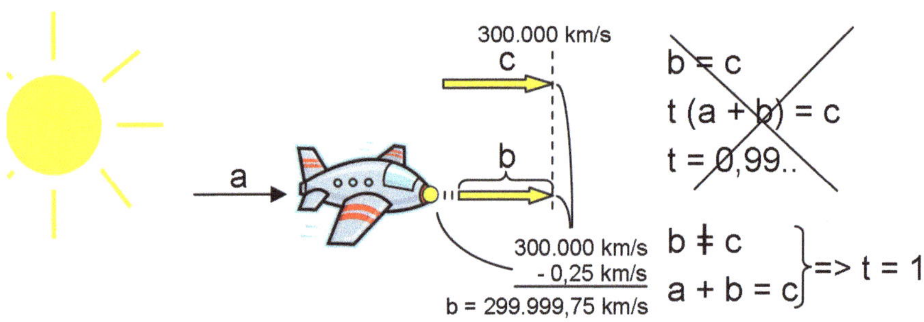

Einstein's General Theory of Relativity

Since an increase in velocity and gravity are experienced the same, Einstein decided to generalize the above theory and published his **'General Theory of Relativity'** in 1916. For that reason, this theory states that the stronger the gravity the slower the time. Or that time would pass faster on the moon. *Any measurable time difference could possibly be explained by the following: in parallel with an increased gravity also the number of bound string chains increases, and possibly also the string density which surrounds the time-determining vibrations of the atoms in the atomic clock increases.* It is not completely surprising that this argument about slowing down of time (time dilation) in the relativity theories also requires patches to become water proof. *Without* these patches, all "photons" originating from places such as the sun or Jupiter with a gravity far greater than the Earth's gravity would come from another time, if to follow the said theories. The described time dilation has also thought to be an explanation to the so-called gravitational redshift of light coming from a place with higher gravity than the gravity of the receiving place. However, this can be explained by differences in string density, and can be compared to the sound of a guitar string that changes when the string is released.

Conclusion

The damage, inflicted by Einstein on our understanding of the universe was so extensive that a belief on so-called ether (dark matter) with its strings and string chains became "illicit". The blame belongs primarily to the fantasy-filled explanations by Lorentz on an experiment run by Michelson and Morley in 1887. Stroke's and Planck's theory that matter pulls the ether (strings – string chains) was disapproved by Lorentz using mathematics. However, to disapprove a theory, more is needed than a mathematical calculation that does not fit in. An incorrect assumption can indeed result in selecting a wrong mathematical model. (But in opposite cases, such as calculations of collisions between galaxies, an explanation could be found to a mathematical result. In that case, the mass of dark matter. See Section 4, 'Kinetic energy and string chains'.)

Figure Chaos is based on the failed attempts to find proof to support a logical explanation to the result obtained by Michelson and Morley. Lorentz's questionable theories are the biggest culprit in this construction flub.

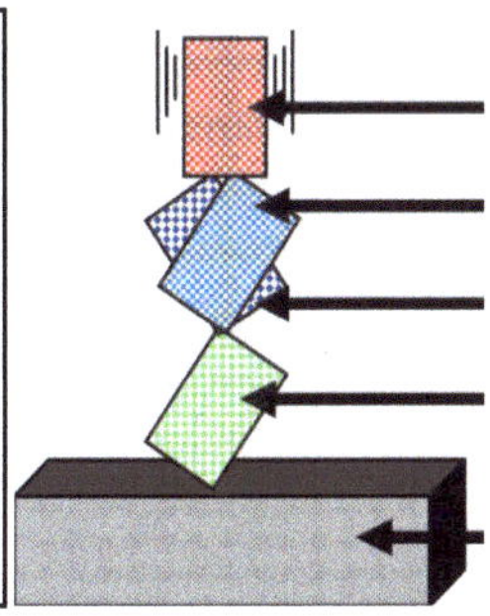

The relativity theories from 1905 and 1916 are based on

Lorentz's theory on time transformation and length contraction from 1904 that was based both on Newton's Galilei transformation

and on the failure to prove that matter can pull ether in an attempt to explain

Michelson's and Morley's experiment completed in 1887.

Death of the photon

The photon has no right to exist

According to the established science, the speed of light is constant in vacuum. It remains so constant that it does not change even if the light is sent from a moving light source. Following the logic, what we define as electromagnetic waves of photons must instead be waves through the media where the light source is located. This can be compared with an object that moves in water. Those pressure waves it sends in all directions have all same speed, both in front of the object and behind it regardless the frequency of the waves. By moving a meter in the waves either to or from the object causing the waves, it is possible to register different frequencies of them. But their speed of expansion remains constant. And like waves through an ocean of strings (ether) moves the light.

A photon's mass is assumed zero when not moving, and it exists only when it moves, which is to admit indirectly that it does not exist. It is just a wave. The wave on the water's surface does not exist either when not moving. A question is also how a so-called photon is able to "bull its way" through the universe's Ocean of Strings – the dark matter – for billions of years without deceleration.

An experiment completed by the Laser Cooling and Trapping Group, the National Institute of Standards and Technology in the USA, a light signal at a speed that was higher than the normal speed of light was successfully generated in 2012. The light signal from a laser was placed on an existing carrier light wave. The result comprised of the existing carrier wave, and a light signal traveling with a speed higher than the normal speed. The result form the experiment can sees as another piece of proof that light consists of waves moving through a media, and not of photons.
See **https://www.nist.gov/news-events/news/2012/04/first-fast-and-faster**

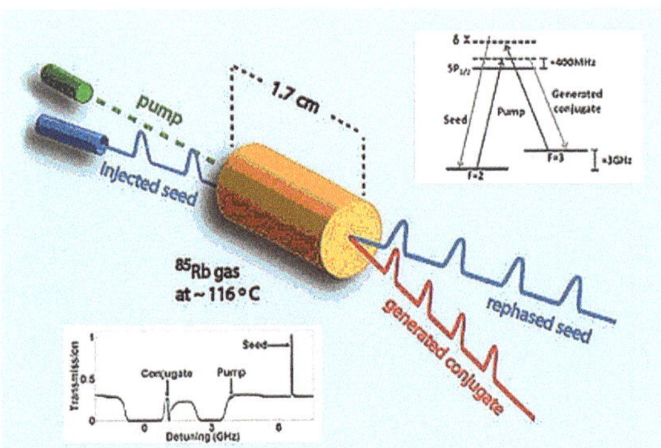

Schematic diagram of the fast light experiment in PML's Laser Cooling and Trapping Group.
Inset at top right shows the rubidium energy levels relevant to the experiment. Inset at bottom left shows the relationship between frequency detuning and gain.
Laser Cooling and Trapping Group researcher Paul Lett and colleagues report in Physical Review Letters that this new method of generating "fast light" has resulted in a pulse that travels up to 50 ns faster over the length of a 1.7-cm cell than it would if it were moving through a vacuum.

Same effect can also be observed when two wave patterns traveling on a water surface in almost the same direction interfere with each other. This results in waves that travel with a speed higher than normal and that spreads over the other wave pattern traveling at a normal speed.

Just like an airplane flying faster than the sound, and boats that can move faster than a wave moving on water, the speed of light should not be the upper limit for matter. Naturally, more energy would be needed, in the above case produced by the plane or boat, to push through the surge of strings which is formed by moving at the said speed. However, and based on how this book explains interference between the subatomic particles and their surrounding string, an atom would be teared apart as soon as such surge begins to form. Outside the ocean(s) of strings housing the universe, no speed limits (and no light) would exists. What this chapter says about light applies naturally to anything that has been considered as radiation of so-called photons.

The so-called photon is just a wave motion.

The argument that light is not a moving particle but just waves moving through string will most likely to be met with following counter arguments: Four hydrogen atoms combined will result in one helium atom plus "photon radiation" (light). And since the weight of a helium atom is lower than the weight of four hydrogen atoms, the explanation to the difference in weight must be that light consists of particles, of so-called photons.

As previously described, gravitation affects the number of string handles in an atom. Study Figure 4 and the figure text. The inner electrons' string handles reaching toward other electrons are shielded inside the core of a helium atom, which provides a fewer available and free string handles for the four hydrogen atoms. This explains the difference in what we understand as weight. The masses should be identical. See more in Section 4, 'Kinetic energy and string chains'.

Light that rounds stars has been shown to bend and get a somewhat changed direction. This has been explained by a gravitational force that is powerful enough to have an impact on the photons. But the explanation should instead be that an increased string density aligns *indirectly* with increased gravity because the string density should be visibly higher around larger clusters of atoms, and not because of the gravity itself. The more atoms, the more subatomic string handles, which consequently results in a higher density among string chains. And all wave patterns traveling through matters with a different density will have their direction bent toward the higher density.

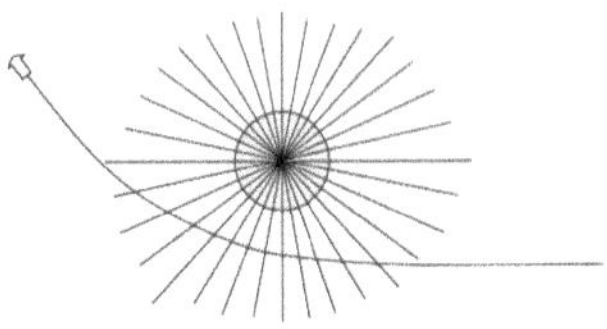

Figure 7. Shows light that travels through a higher string density existing around a star.

The so-called black holes have been considered black because the gravitational force would be too great for the photons, the "light-mediators", to break away. But if light is explained as waves moving through string chains, why are then black holes invisible? One answer can be that the activities taking place in a black hole are not comparable to those taking place in a star. Furthermore, it could be assumed that string chains on the surface of a black hole are so tense/stretched out that waves moving through them create a frequency so high that is cannot be registered, or it cannot be mediated through string chains. Some areas around a black hole could even temporarily lack strings due to the gravitational force thus making it impossible for any wave movement or gravity to be mediated through them. But the light that has been observed shimmering around something that could be a black hole could originate from shock waves created by strings being pressed out. See 'Big Bang Could Never Have Been' in Section 5.

More nails in the photon's coffin

Thanks to the blinkers worn by the trained high priests – who in their collective anxiety sacrifice their common sense on Einstein's altar in famous scholar institutions, strongholds of arrogance – the risk of being publicly executed/burned at a stake is minimal if you argue against the official line that the photon does not exist. One hundred years have the said priesthood enjoyed a blanket of security in its believe in existence of the photon.

It is easy to understand how the old Greek belief in the four elements, earth, fire, air and water, got established. It aligned with what could be observed. But for two thousand years, this theory has needed countless excuses and alternative explanations to survive. Believing in the photon follows the same pattern and new theories have been needed to explain what cannot be explained, such as a photon somehow would know what another photon will do. This is an example of a myth that is born out of a need to patch together an incorrect understanding of the reality. In 2000, Yoon-Ho Kim et el. conducted an experiment that keeps this myth alive. The experiment (2011) is described and available on the internet. See **http://en.wikipedia.org/wiki/Delayed_choice_quantum_eraser**

The puzzle studied in the experiment has a more logical explanation if the phenomena is thought as strings moving in wave patterns with different amplitudes, and two waves sent out at the same time naturally display same amplitude. If the amplitude is not forceful enough, neither the upper not the lower wave pattern is capable to penetrate the lens or the semi-reflecting light. Study Figure 8 and the figure text.

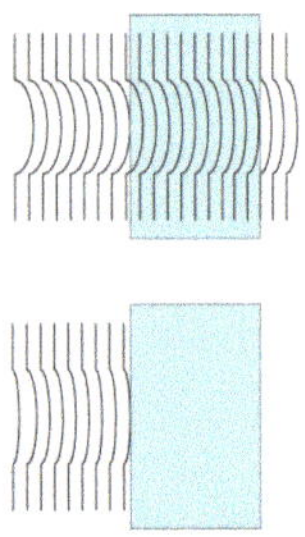

Figure 8. Shows two different string chains moving in a wave pattern toward a glass. Amplitude of the lower wave is too small to impact the string chains in the matter, and are reflected.

The upper wave movement "makes" it through.

Another suggested explanation to a photon's wave patters is the energy created by the photon. However, a milieu is needed in the space that the said photon can interfere with in order for such an energy field to be generated. Strings? A photon cartwheeling in relation to its own counter weight could be able to create a wave. But a photon with a counter weight is also far-fetched.

What is understood as an electromagnetic wave pattern generated by the photon is actually nothing more than waves traveling through string chains. The moving string chains move/let go off electrons they are attached to. And electrons that come off generate electromagnetic effects mediated as waves through string chains.

In a documented experiment, a laser beam was sent through a gas that had been cooled down to a temperature close to the absolute freezing point. It was visibly obvious that the light traveled slower in the gas. Atoms at this low temperature move only minimally, and the inter-electron string chains described in this book are hence maximally rigid with minimal interaction with external string chains. Subsequently, a wave cannot pass through these string chains as easily as it otherwise would. Therefore, the experiment referred above could be one more nail in the photon's coffin.

This can be an excellent moment to underline the difference between the expansive movement of light and magnetism. Light can spread out as waves through separate strings/string chains while a magnetic field spreads out relying of linking together string chains as already explained under ‘Magnetism’.

What could be used to support the photon theory is that light can be polarized. For such phenomena, it is speculated that a polaroid filter lets through only photons that have a vertical wave pattern. But one can equally speculate that what is filtered out are waves moving through string chains so that only the waves moving through, for example, vertical string chains, in other words, the vertical wave pattern, are let through. Polarizing characteristics of a material reveal a lot about how its outer surface is constructed, and it is easy to assume that it contains directionally similar string chains of a crystalline structure for example, that can only forward waves parallel to them.

Upon examination of gaseous chemical elements, they can be illuminated with a light that contains all colors, in other words, normal light. The illuminated gas will then be impenetrable to one or more of the wave lengths (colors) that the light contains. The gas can then be identified by its **spectral lines** (black stripes) created where no color for a certain wave length is visible in the spectrum. The factum that chemical elements can absorb selectively only certain wave lengths, and emit /reflect others, indicates strongly that light cannot consist of particles. **It is reasonable to think that *only* waves** (see Figure 9a and 9b) **can hit so accurately the electrons or their string chains.** And each string chain has naturally its specific resonance frequency, which explains the different colors.

No more nails should be needed on the photon’s coffin. Who will attend the funeral? It is unlikely that they come from Chalmers Technical University because there they believe that even virtual (!) photons exist based on an experiment. It is *just* another example of a crumbling model of the reality being amended with a patch. See ‘Casimir effect’ in Section 4.

Possible explanations to phenomena within quantum physics

In an attempt to better understand certain behavior within an atom, numerous theories have been developed during the past few hundred years in a field called quantum physics. There is a reason to believe that these behaviors can be explained by what has already been described here, namely the dynamic condition between the subatomic particles and strings. See Figure 4 describing the cooperation and interaction between a proton, neutron, electron and strings. This example uses a helium atom which is the simplest atom including all the mentioned ingredients

The result of the described interaction and interaction between the kinetic energy of electrons and string chains, when electrons are detached or attached, should produce effects that one tries to understand with the theories of quantum physics, such as e.g. location of electrons (shell) and their bound energy (type; s, p, d, etc.).

An electron's kinetic energy must become gradually variable depending on whether it detaches from one, two or three string chains. How tightly coupled the nearest electrons are should also affect the energy of the detached electron. Based on the string handles described in this book, one could build a model of an element, see how the electrons are connected to each other, and thus gain a mapping and understanding of the element's properties.

See figure 9a and 9b. The energy released by the handle exchange of string chains to the subatomic particles should be one, or a multiple, of what has been defined as a quantum, that is, a quantum for each loosened string chain. These differences in the energy levels of the electrons should explain differences in the color emitted from different illuminated objects. Blue, which has a higher frequency, is therefore more energetic than a red color. (Illumination, or some other form of irradiation, could perhaps also be applied to the tugs of the string chains that make up the gravitational forces.)

Because of all the movements in the Ocean of Strings, the subatomic particles are continuously exposed to string chain handles moving from linkage to linkage. It is hence normal from a statistical point of view that an atom comes loose before it has time to catch the next handle. This familiar phenomenon is known as **tunnel effect**.

Quantum physics is also trying to find solution to unexplainable patterns observed in the photon's behavior, but as already pointed out, the photon cannot be anything but a wave traveling through strings. These strings, while moving in a wave-like pattern can together with the subatomic particles have an impact which traditionally has been assigned to the photons.

Figure 9a. Shows **waves traveling through string chains causing an electron to detach** from one or more other string chains (not shown in the figure). At the same time, we can see the waves to a stop.

Figur 9b. Shows an **electron whose movement produces a wave motion through string chains**, when the string chains that held it (not shown in the figure) break. The generated wave motion can be visible light.

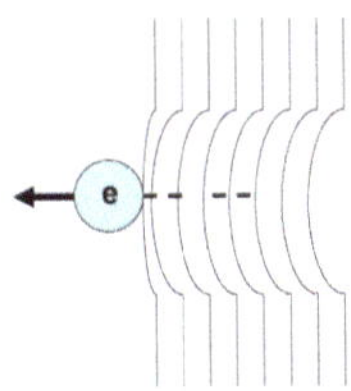

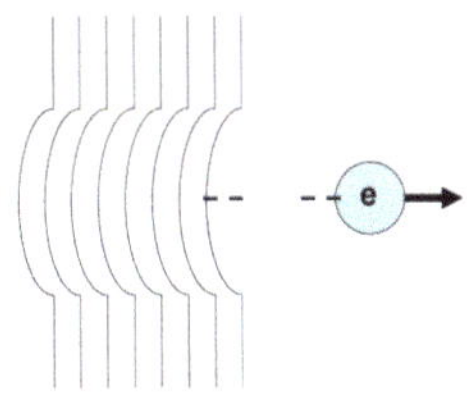

Wave-particle duality and the double-slit experiment.

Wave-particle duality refers to a subatomic particle also exhibiting properties of a wave motion. To study both this and the properties of light, the double-slit experiment is used.

But the fact that you can see effects of wave motions from "ejected" subatomic particles can be explained by the fact that they are moved through dark matter in which the measured wave motions are created. And light, as this book claims, is not a particle, but only wave motions through dark matter. As long as these facts are not accepted, nor can the result from the experiment be explained.

Section 3

Summary and Evidence

The logic indicates the following:

Strings are more than just building material to subatomic particles. Strings form the entire "ocean" that surrounds and penetrates the universe. In addition, there is an ongoing interaction between all subatomic particles and the strings surrounding them which in their turn are directly or indirectly in contact with everything.

Interlinking of the strings to form chains, and them interlinking between subatomic particles explains magnetism, gravity, and the strong and weak nuclear force, in other words, all the forces that are described to be the most basic forces in physics. The strings mediate the gravitational force but are not target to its compressing powers. However, the string density increases with existence of matter.

Any radiation, including light, considered to be photon-based, consists of waves traveling through the Ocean of Strings, which the universe is engulfed by and which the universe consists of.

The evidence: helium II

Zero gravity, super liquid, heat conductors, Cooper pair and superconductor.

Well-documented experiments demonstrating interesting characteristics of helium II seem to provide evidence supporting the logic of this book. Helium II is liquid helium at a temperature close to the absolute freezing point. It consists, at least partially, of He_2 molecules with a lifespan of only 13 seconds. In one experiment, this helium can be observed to acts like a never-ending fountain, and in another, its surface tension seems not to be affected by gravity. The liquid crawls freely up along the walls of a jar. One explanation could be that the electrons in a He_2 molecule have only 0-4 available string handles which would normally connect to string chains mediating external gravitational force. He_2 has eventually up to two double string chains between the electrons, as shown in Figure 10, and subsequently no free string handles. The results in the gravitational force failing to keep down the liquid when its molecules hit against and other and begin to crawl upwards.

The two images below depict the experiment. See http://www.youtube.com/watch?v=2Z6UJbwxBZI

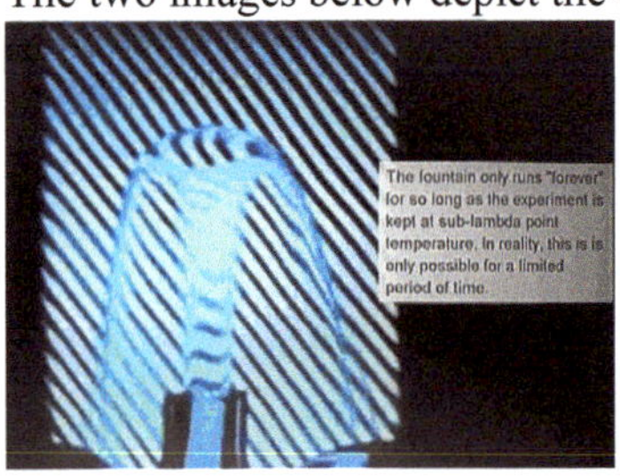

Helium II is described as **super liquid**, a liquid with zero viscosity. This could be explained by a complete absence of available string handles that normally function as a cohesive linkage between the molecules, also known as Van der Waals force. It can be of interest to note here that according to this book Van der Waals force and gravitational force are two sides of one and a same thing. Helium II **possesses a superior heat-conducting capacity** – 170,000 times better than that of water – that can be explained by kinetic energy that is unhindered by any surrounding string chains.

Certain chemical elements have none or extremely little electric resistance near the absolute freezing point and up to 100 degree Kelvin, characteristic to **superconductors**. The currently accepted explanation is that the electrons build a so-called **"Cooper pair"**. This book argues that such Cooper pairs are **two electrons bound together by more than one string chain.** In other words, they have at least one *extra* string chain that can mediate bypassing electrons without binding them to the atom/molecule in question. This is why a flow of electrons in an electrical current can flow freely without resistance in such a material. And since electrons passing by cannot attach to these atoms/molecules we will not get the electrical polarization in the material that would normally generated magnetism slowing down the current. The more string chains the material contain, the more current can pass without resistance. The explanation of the inter-electron string chains provided by this book should be an important contribution to better understanding a superconductor. This should help in designing molecules with electrons that are linked together by double string chains at higher temperatures than needed for the current superconductors. Therefore, helium II with its double string chains should be a superconductor as explained in the report as stated in the book. And correctly so; helium II has shown to be a superconductor in addition to the above-specified characteristics. (Superconductor characteristics are also described in 'Electromagnetism and electricity' in Section 1.)

Figure 10. Shows a He_2 molecule (helium II) with two double bonds between electrons, and subsequently without any available string handles, for "mediating" gravity. Even in those cases where the molecule has only one or no double bindings, it still has unusually few available string handles, which should mean a lower weight than the combined weight of two separate helium atoms.

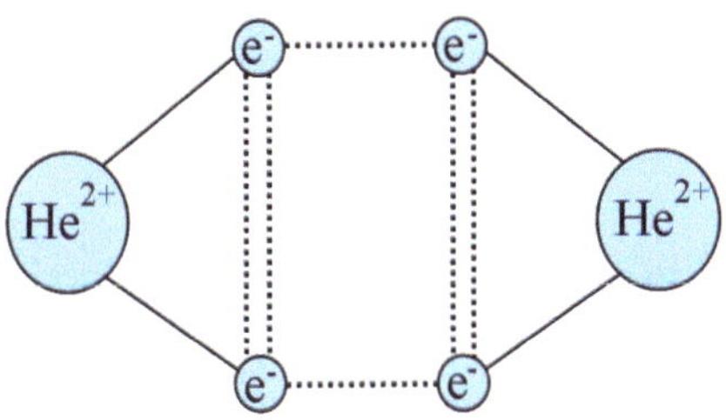

Important!

1. With the description provided in this book on string chains that exists between subatomic particles as the basis, it could **first** be stated that a molecule consisting of 2 helium atoms, if it exists, should be almost weightless.

2. Descriptions of helium II was **subsequently** found on the internet. See the following web site (available in 2014): **http://www.youtube.com/watch?v=2Z6UJbwxBZI** Observe the expression **"It appears to defy gravity!"** in the video.

3. The web site is a proof that the **theories in this book is an accurate description of reality**!

String chains as a concrete reality is proved by the frog that lay on these, as in a hammock, in a very strong magnetic field. This was described in Section 1, under ""Magnetic" repulsion does not exist"

The chase for the nullifier for gravitational force continues

Some of the conclusions have already been summarized in Section 1 under the title 'Otron plate; String handles and intersecting magnetic field'. The knowledge that gravity is mediated by string chains provides us great possibilities. With this in mind, we should be able to use helium II for any object we wish to make weightless because helium II does not mediate string chains. But first, a method to store stable helium II must be developed.

It might also be possible to break the gravity-mediating string chains, or to prevent them from attaching to the subatomic particles in the object which is subject to the anti-gravity procedure, by using radiation with different frequencies possibly polarized in a certain angle.

Section 4

Some Applications Based on the Conclusions of This Book

Kinetic energy and string chains

You should read 'The constant speed of light, and Einstein' in Section 2 before reading through this topic. The classic mechanism has the following equation for kinetic energy: $E = M * H^2$. It says that an object's kinetic energy increases by a velocity squared, and it was established for about 300 years ago by Leibniz among others. The explanation to the increase in kinetic energy by a velocity squared is: When a velocity of an object is increased the kinetic energy of what already has been established at a certain velocity increases also, and can be described as follows: $(M * H) * H = M * H * H = M * H^2$. Later, the factor 0.5 was added, and the new equation looks like this: $E = 0.5(M * H^2)$, i.e., $\mathbf{E = 0.5 * M * H^2}$.

It is easy to demonstrate with a simple test that the kinetic energy in a car is not linear in relation to its velocity. It is done by letting a car roll with transmission on neutral on a flat surface to see how far does it roll before it slows down from a higher initial speed than from a lower initial speed. Drive for example 90 km/h and let the car roll until the speed reached 80 km/h. Compare this to the distance your car travels if you drive 30 km/h and let the car role until it rolls at 20 km/h. Despite a much higher aerodynamic drag, the car will roll longer at higher speed.

Leibniz discovered the interesting factum that an object must store more energy when the speed is increased from 80 to 90 km/h than from 20 to 30 km/h. The above equation is based on hands-on experiments. One interpretation of the results obtained from such tests can be that an object's mass increases every time its speed increases, and velocity squared reflects this increase. To understand the difference between mass and weight think of an object's mass being the same as its weight on the Earth's surface when the object's relative speed in relation to it is zero. The weight remains constant and can be easily demonstrated by weighing the object at different speed. However, the weight changes depending on the size of gravity.

An increase of the mass by a velocity squared means that the faster an object travels the heavier it is to increase its speed, even with no aerodynamic drag and friction. "Is a body's inertia dependent on the energy it contains?" was one of Einstein's questions in his study published right after the "Special Theory of Relativity". In this study, he found a relation between the energy (L) of the particles that radiate at a speed of light (C) and the emitted mass (M). This is described by the following equation: $\mathbf{M = L / C^2}$. 'L' in the equation is thought to stand for "Lorentz", whose work Einstein had based his ideas on. The equation was later reformulated to the famous; $\mathbf{E = M * C^2}$. If 'E' was not his own initial, one could suspect that Einstein had had his eyes on the well-known $E = 0.5 * M * H^2$ from the very beginning.

To transform energy to mass was most likely never the intended idea with the formula. It is also important to understand in this context that the particles emitted from the mass do not disappear. On the contrary, the equation can be considered as an expression for the so-called resting energy for matter, i.e., the energy needed to send out subatomic particles, in other words, it represents the inertia (E / C^2) for the mass (M). And the formula $E = M * C^2$ shows the energy, such as the amount of joule, that is produced at the loss of the mass (M).

Another interpretation of the equation can state that if the mass increases, energy increases, too. The speed (of light), as already stated, is constant. Calculations within the field of particle physics has assumed that mass radiating out consists of the so-called photons. The measurable loss of weight however is explained in the last chapter under Figure 4 of Section 1. **Kinetic energy at atomic and subatomic level originating from the described activities is mediated as waves, such as gamma radiation, through the surrounding "Ocean of Strings"**. To conclude about the photon's right to exist; a photon's mass is assumed zero when not moving, and it exists only when it moves, which is to admit that it does not exist. Likewise, a wave on a lake exists **only** when it moves. And **only** one wave has one, per medium, constant speed.

We return now to the example about inertia and the car. The popular explanation mentioned shortly in the introductory part has been that the energy already built up is what needs to be given more energy. But it is unfortunately both illogical and difficult to comprehend that an object's kinetic energy is not linear in relation to its velocity, because **the velocity is without question always relative.** Therefore, a look further into the horizon is needed to find other and more radical theories. Let's argue that an object's mass does not increase but it carries an invisible mass, also while moving at lower speed. The argument would be:

The faster a body travels, the more strings/string chains gets latched onto it. This does not affect its weight but its mass increases. And strings, the building blocks of everything, are the origin of all mass.

Evidence for the argument could be found in the phenomena observed at studies of colliding galaxies. One of conclusions on these studies has been that dark matter (strings) tags along as extra mass. See Figure 11a and 11b. Kinetic energy includes also so-called dark matter, hence lacking the linearity in relation to the matter *only* and its velocity. String chains have the same velocity as their surrounding because they are attached to it. An object moving at an increasing speed catches strings and string chains that subsequently become detached from their original surrounding. Strings that are pushed and attached to the object's atoms will contribute to a portion of the kinetic energy by their mass. The higher the speed, the more strings get attached and secured. As previously mention, Jean Fresnel assumed already in 1818 that we are not only being surrounded by "ether" (dark matter), but it also penetrates the matter. Even assumptions that moving matter takes along ether, were heard. This book argues that this is the case, in other words, the ether (strings and string chains) both penetrate the matter and are dragged along by it.

A proof that the above statement is a correct description of the reality can be found in that it is not heavier to move against gravity on the side of the Earth that happens to be turned toward the direction the Earth moves around the sun.

If, based on the above, the following is assumed:
1. An object (the "visible" mass that can be weighed) increases its kinetic energy in a linear fashion with its velocity.
2. And an accumulation of dark matter (strings and string chains) increases as a function of the object's weight and its velocity.

Then the following equation for kinetic energy can be written using internationally better known symbols: $\mathbf{E = M * V + M_D * V}$, where E = kinetic energy, M = visible mass, V = velocity, M_D = accumulated dark matter (Matter Dark).
This formula should then equals $\mathbf{E = 0{,}5 * M * H^2}$. That is; $\mathbf{M * V + M_D * V = 0{,}5 * M * H^2}$.

That equation gives us the mass of the accumulated dark matter: $\mathbf{M_D = (0{,}5 * M * V) - M}$.
Inserted in the new formula gives us: $\mathbf{E = M * V + ((0{,}5 * M * V) - M) * V}$.

If we test the equation for an object that weights 10 kg and travels at a speed of 10 m/sec, we get following:

$$10 * 10 + ((0,5 * 10 * 10) - 10) * 10 = E \rightarrow 100 + ((0.5 * 100) - 10) * 10 = E \rightarrow$$
$$\rightarrow 100 + (50 - 10) * 10 = E \rightarrow 100 + 40 * 10 = E \rightarrow 100 + 400 = E \rightarrow E = 500$$

The formula can be reduced to the old equation for kinetic energy, and it will therefore provide the same result but **we can now calculate also the amount of dark matter that follows each object at a certain velocity**. For example, the amount of dark matter is 40 kg for an object that weighs 10 kg at 10 m/sec, and 240 kg for the same object at 50 m/sec. But because this equation show a negative kinetic energy for the dark matter at velocity less than 2 the constant '0.5' must be replaced with a variable that successively changes from 0.5 to 1 when the speed changes from 2 to 0. This could also describe how the interaction of matter with dark matter changes when the speed get closer to 0.

It would be meaningless to reformulate the equation (Einstein's) $E = M * C^2$ in order to observe dark matter because it would only describe the energy for a wave pattern at a constant velocity through dark matter. The reason why Einstein got his change was, as mentioned earlier, a battle field left open by the discussions about an explanation to Michelson's and Morley's experiment. The outcome was, as we know, to exclude the possibility that dark matter would be dragged along by (visible) matter. Yet, equation $E = M * V + ((0.5 * M * V) - M) * V$ together with such explanations why it is not heavier to move against gravity on the Earth moving along the direction of the sun, shows that dark matter binds to matter. It seems that Stokes and Planck were right, and that Lorentz's theories, further developed by Einstein in his relativity theories, lay on a shaky ground. See also Figure 6b in Section 1.

Figure 11a.

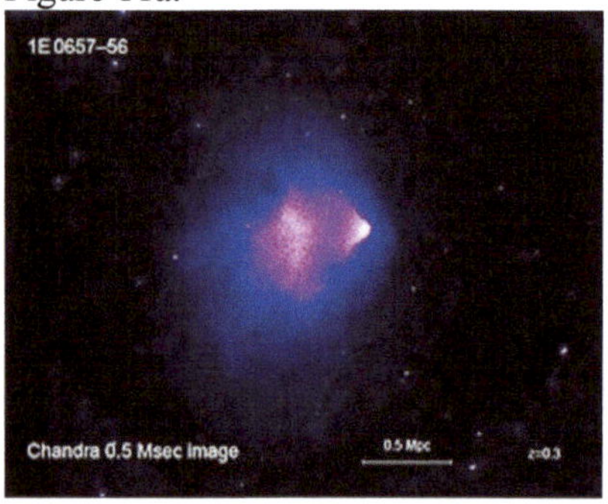

Figure 11b.

On left (Figure 11a): a photo of Bullet Cluster taken by Chandra X-ray Observatory.
On right (Figure 11b): shows an image of MAC J0025. The image is a composite image of a photo taken by the Hubble telescope and data from Chandra's ACIS detector. According to the Space Telescope Science Institute, the image shows that a gigantic collision has separated dark matter from normal matter, and that this is an independent proof of a similar effects detected also by Bullet Cluster (Figure 11a). The detected dark matter is colored blue in the image.

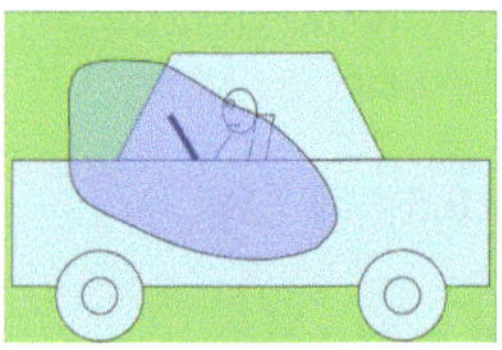

Figure 11c. Dark matter is luckily not so dark.

If it was possible to "cut off" the gravity-mediating string chains connecting electrons of separate atoms, it would minimize or eliminate completely any dark matter being captured. It would also explain observations of so-called UFOs with their impressing acceleration because 'MD * V' would be eliminated from the above equation, and we would get: $\mathbf{E = M * V}$. In other words, it would not become heavier to increase velocity once it has already been increased.

The neutrino, does it exist?

The official word is that simultaneously with a positron leaving a quark something that can be registered is emitted. This something was named neutrino, a particle that can cross the universe and travel through mountains. *But if we accept that we exist in an ocean of strings it would subsequently be difficult to accept that particles, such as neutrinos and photos, could travel through these oceans without ever being slowed down.* As previously explained, and as argued in this book, the so-called photon is nothing more than a wave that carries a resonance in string chains of the type that can bind electrons to each other. In a similar manner, the so-called neutrino could be just a wave traveling through string chains (positrons, see below) of the type that can bind electrons to protons. This type of wave pattern would be created when such a string chain becomes detached from a proton, and hence becomes a wave pattern harmonized with other positrons. When talking about formation of waves, small atoms are indeed the mediators of waves traveling at (almost) this level across the entire planet after, for example, seismic activity. It should be pointed out that both the neutrino and the positron appear to be particles that do not get stuck in string chains and therefore do not bind to other particles via these either.

And what is a positron?

Is it a “conventional” particle, or is it just a string chain between a proton and an electron? Where to draw a line between a small ball of strings, i.e., a particle, and a chain of strings? If a positron would be part of antimatter, would it be connected to an electron that exist in a neutron? If the quarks of a nucleon are continuously moving, and one of the positrons is attached more securely than an electron, then this would result in a hydrogen atom instead when the detached electron becomes captured. This explains the short life span of possible antimatter. **One of the theories presented in this book is that a neutron does not contain any positrons. This sound logical but can be incorrect.** A neutron emitting an electron and turning into a proton could be explained by a positron that can eliminate the electron.

Figure 12.

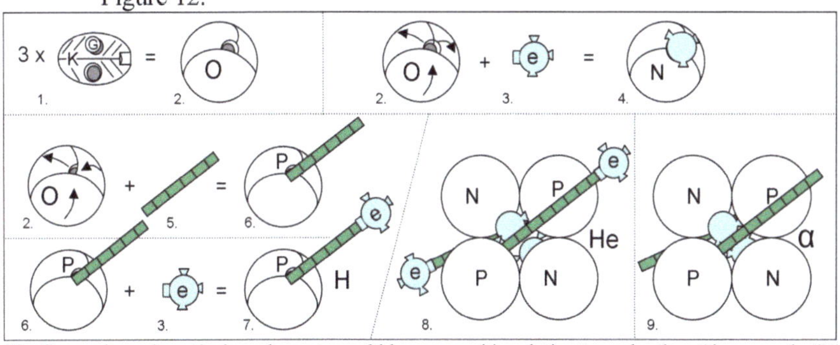

The figure shows how the key elements could be arranged in relation to each other. Three quarks (K) with gluons (G) form together (1) the “baffling” otron (2), the nucleon which both the neutron and proton are based on. An electron (3) together with the otron form a neutron (4). If the otron attaches to a string chain (5) instead that can form a connection to an electron, then the otron becomes a proton (6). Subsequently and based on logical thinking, the above string chain (5) should be the positron. Depending on which way the quarks are turned they will either fit an electron (one up and two down) or a positron (two up and one down). An otron's coupling is hence dependent on what it is connected to. If an otron can exist as a separate entity it could be a so-called “**WIMP**”. See ‘Parallel worlds’ in Section 5. A proton (6) together with an electron (3) becomes a hydrogen atom (7). Four hydrogen atoms make, as we know, 1 helium atom (8) with its nucleons bound to each other with the help of 2 “inner” electrons. Along this fusion, 2 positrons are lost. It must mean that 2 of the quarks have turned. See also the neutrino above. Is the neutrino a released positron, or is it waves following the positron moving through the Ocean of Strings? Many atomic nuclei contain alpha particles (9). If a loose alpha particle captures 2 electrons it becomes a helium atom, as shown in the figure.

Casimir Effect

In 1948, a Dutch physicist Hendrik Casimir and his countryman Dirk Polder, while studying the Van der Waals force, observed a force between two plates in a vacuum. The force is neither attracting nor repelling. The force can be measured and it increases substantially with decreasing distance. For example, it is assumed to be about 1 kg/cm^2 at attraction when the distance has decreased to one hundred thousandth of a millimeter using vacuum in the experiment guaranteed that air pressure would not affect measurements of this force which was named 'Casimir Effect'.

In 2011, scientists at Chalmers succeeded to demonstrate the particles, virtual (!) photons, that were thought to explain the Casimir Effect. The "photons" that were produced with the help of a magnetic field alternating its direction at a high frequency were thought to originate from the virtual photons. Normally photons are believed to emerge when electrons move to another shell – but there are not atoms in a vacuum. This was the reason why virtual photons were believed to account for the force within the Casimir Effect, unclear how it was thought to work. This book has "killed" the photons, though, and the peculiar idea about virtual photons with no attachment to logical reasoning, only demonstrates that the further away from the truth the more farfetched are the explanations.

As already clearly shown and explained, everything previously thought as photons is nothing more than waves traveling through string chains in the ocean of strings that surrounds us. Anyone who has read this book will understand easily how a magnetic field manages to create these waves. When there is reason to believe in virtual particles, they could possibly be thought of as so-called super symmetric particles consisting of other strings, and/or other string combinations than those particles that build up our world. See more of this topic in "Parallel worlds" in Section 5.

Figure 13a.

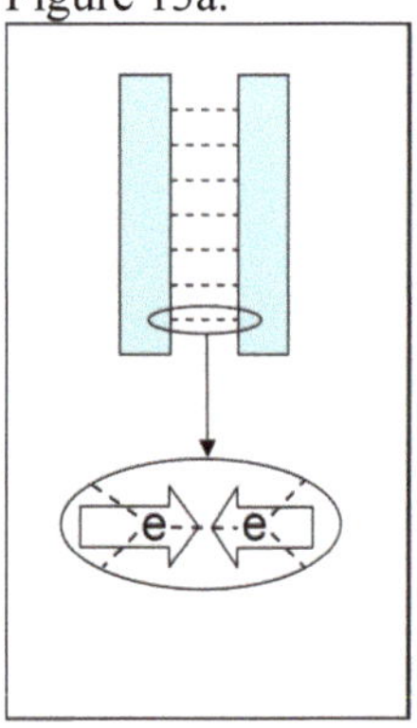

Figure 13b.

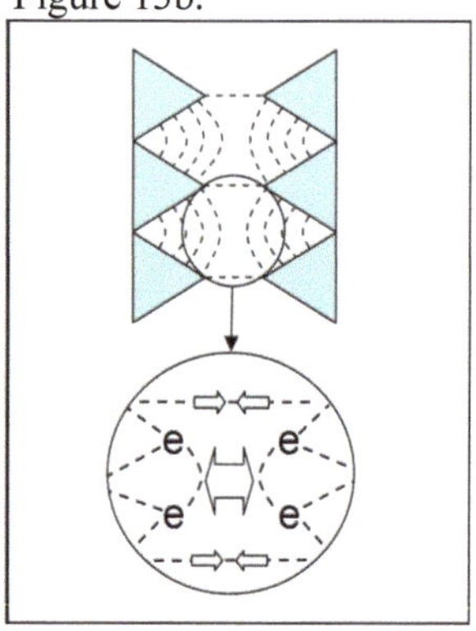

The above figure illustrates the Casimir Effect in a way as it is described in this book with string handles located between electrons. See Figure 5 for string handles on electrons (e = electron) Figure 13a shows that Van der Waals attraction between two objects should be identical, such as described in this book. Figure 13b shows that repulsion between two objects with a special design or positioning should be identical, such as described for magnetic repulsion in this book. In addition to repulsion, there should also be attraction although it is weaker than the repulsion. Both forces are most likely present also when attraction is registered.

Section 5

The Universe – Ocean of Strings and Philosophy

Parallel worlds

Are there possibly other classes of chemical elements that are built of building blocks (i.e., strings) that are not “our” atoms. The purpose with the thought experiment is to demonstrate to the reader that there should be more than what we can see and touch. Concepts, such as supersymmetry, dark matter and WIMP are indeed developed based on recognized experiments from established institutions.

The term ‘atom’ got its name in Ancient Greece. At that time, it was thought as the smallest building block, and therefore indivisible. (“Undivided” on Creek = “atomos”). As evident in the book, the atom has during the past roughly over one hundred years had to “open up” to reveal its innermost secrets. Ever smaller elements have been discovered step-by-step. The book has moved one step further, and landed on the string level. However, it is not entirely novel thinking that subatomic particles would possibly consist of something called strings. Claims about relatively large distances that might exist between strings have also been brought up. Therefore, it is not hard to understand how someone can be misled to believe that the universe could be pressed together to a primal atom preceding the so-called Big Bang. But as becomes evident in the book, all our subatomic particles are built up of strings without inter-particle ranges and possible associated “mysterious” force fields.

It seems logical that a complete atom consists of a limited number of string types. One or more types of strings, or their combinations, form electrons, others form protons, neutrons, etc. Vi can refer to these strings as “Class 1” strings. (In this class, there must be more than one string, otherwise only be one type of subatomic particles would be formed.) The designation “group” should be avoided because this word is closely associated with division of atoms into groups in the periodic table used in chemistry.

The existence of different types of strings can mean that the strings are not the smallest building blocks! However, this is a different story and will not be further discussed here.

It is highly likely that in addition to “Class 1” strings, there are others because it is unlikely that to construct “our” atoms all string types in the universe would be needed. Non-Class 1 strings should not react on any of the strings that build up our world. Otherwise, we would indeed be categorizing them into Class 1. Relying on logic, we advance to assume that these non-Class 1 strings could with high probability build their own constellations (atoms or similar), and further, that Class 2 strings and Class 3 strings, etc., may build their own worlds. The established theory of **supersymmetry** argues that our subatomic particles have equivalent particles inside other dimensions.

An assumed molecule composed of Class 2 strings should thus be able to pass straight through one of our Class 1 atoms because the strings of different classes (according to the definition) do not interact. In other words, we could have parallel worlds in the midst of us. Whether biological life has evolved in such a world is another question. If not in Class 2, then maybe in one or some of the others. A potential creature from such a world would therefore be able to wander through walls, or be exposed to a nuclear bomb explosion without noticing a thing, and even maybe spend time wondering about these same questions.

However, it is not yet possible to show existence of any other strings than those of the class that forms "our" atoms because our instruments and tools are not built to detect something that is composed of other types of strings. But the book has also explained how our subatomic particles interact with each other in the Ocean of Strings through their string handles and chains of strings which could be broken by particles or string chains belonging to parallel worlds. It should also be possible for these "ghost particles" result in direct hits, or stay in a way for our particles. In these cases, "our" particles should be able to indirectly mediate the detection of the ghost particles.

Because light/radio waves are not so-called photons, as argued in this book, but undulate through strings and string chains, makes this interesting. A such a ghost particle or object of a parallel world could hence through its mere physical presence disturb the said wave pattern and simultaneously stay under the radar. To be observed (directly), the said object must create waves, as "our" own electrons do while moving from one shell to another. However, it is not certain that waves from a parallel world can become harmonized with "our" strings-string chains and interact with them.

To conclude: If the assumption that an atom is made of strings is correct, then it is equally correct, logical and natural to assume that there are strings that are not building blocks of the atoms in *our world.*

The reader is invited to speculate on the above theories. Many supernatural events could indeed be explained by the presence of a parallel world.

And as mentioned previously, the universe is filled with something that is called for "**dark matter**" and "**dark energy**". (The term "dark" denotes the invisible nature of this matter.) What this book refers as "strings" is the dark matter, and subsequently dark energy should be movements of the strings.

Lack of evidence that strings exist has been used as an argument against this book. But astronomers who have studied colliding galaxies believe that dark matter (strings) explains some of their observations. See Figure 11a and Figure 11b and the associated figure text in Section 4. Furthermore, scientists have during the past couple of years tried to find proof for existence of **WIMP**s (Weakly Interacting Massive Particles). These particles could be the same as described here. See also Figure 12 in Section 4.

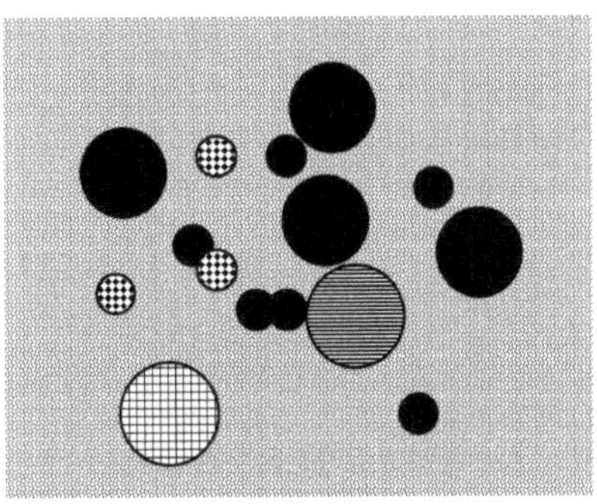

Figure 14. Shows clusters of strings, such as atoms of "our world", "floating inside" the Ocean of Strings.

Big Bang Could Never Have Been

Why do so many people believe in "Big Bang" and not in "Steady State"? Could it be that common sense has been replaced with a blind belief in those who are the experts in the field? In Medieval times, the "experts" believed that the universe surrounded us in its center. And that is what they still believe in. This time by stating that we are located from the universe's edge at a distance equal in every direction. In other words; that we happen to be exactly in the location where Big Bang took place. Sensible scientists are unfortunately not heard, when fanciful claims are promoted. The human intelligence may be great but what is it good for if our judgment has not gotten any better. Refer to "Intelligence – Knowledge ..." at the end of the book.

Two Theories

There have been two competing theories out there about the universe. The older of the two claims, like it appears from its name "**Steady State**", that the universe remains at equilibrium. The theory explains that the steady density of the universe - despite its expansion - is acquired by hydrogen emerging from "empty nothingness". This theory coincides with the content of this book. The later and best known theory, "**Big Bang**", argues that all matter in the universe originate from just one "primal atom" that exploded. Those favoring Big Bang use the expanding universe and the cosmic background radiation to support their argument. They assume that the radiation originates from the beginning of the time right after the explosion. And subsequently, because this radiation has been measured to be more or less equal from all directions, we are "as usual", and just as the priests said in the Middle Ages, in the center of the universe!

When analyzing the constituent colors of the light from a galaxy, it can be concluded that this spectrum is shifted in one direction or the other, depending on whether the galaxy is moving towards or away from us. The analyses, which have established that galaxies are moving away from us, have been taken as evidence for the Big Bang. **But how do we know that the universe does not rotate, and that it is thus subject to a centrifugal force.** A recent discovery also shows that there are some very large galaxy clusters at such great distances, and with locations in various concave shapes that, according to the discoverers, contradict the Big Bang theory. And regarding the findings that have been made about light being affected by gravity, it should be remembered that light is a wave motion through dark matter. Since this dark matter also mediates gravity, its density should vary and thus bend the light accordingly.

Cosmic background radiation.

As described in this book, the radiation ought to be the noise produced by strings/string chains that fill up the universe, almost like dust you can see hanging in the air in a sunny room. Here and there, they get stuck on each other to form dust bunnies, or as in our case, subatomic particles that then form hydrogen atoms. See the introductory section "An introductory summary..." of the book. The noise originating from this activity – that takes place on an enormous scale – should generate radiation. The fact that a growing number of details on the Big Bang theory modeled by scientists is based on small variations in this radiation does not make things any better but can rather be seen as grandstanding. Given that gravity in the universe is mediated by the strings of dark matter, as explained in this book, said variations should vary with the presence and strengths of the different gravitational fields between stars and galaxies. As it stands, the risk for populism cannot be ruled out in the fight for survival among scientists. And people seem always wanting to believe in the most fascinating, anyways.

The Matter in Dark Matter

The universe is, as previously described in the book, an "ocean" of strings. It is endlessly vast and endlessly deep. It has been that way forever and ever, and its infinite eternity can fill an observer with emotions of religious nature. And in that ocean, we exist. An ocean, that by definition in the endless and enormous universe, has no outside and can never be observed from outside in. **This Ocean of Dark Matter has "islands" in which the strings of dark matter have formed atoms that in turn have formed stars, galaxies and even black holes.**

Black Holes and Their Dynamics

Black holes signify a never-ending cycle of death and re-birth for parts of the universe. Relatively stable black holes can be found in the middle of galaxies where stars revolve around them, similar to planets revolving around a star. Yet, no heavier "hub" is needed to hold stars together. Twin stars are an example of this. Depending on the closeness and direction of the movement, matter is providing "food" to the black hole. It is first when the hole has pulled enough material to crush atoms to strings with the gravitation when the hole turns black. **And a black hole is black because the crushed atoms no longer are able to generate photoelectronic, chemical or nuclear processes.** Strings with no connection to atoms are not impacted by gravity at all, and will be pressed out. This should explain the phenomena called **quasars**, and maybe also so-called **gamma-ray flashes**. See Figure 15. This further decreases the credibility of the existence of photons. In the way they have been described, they would indeed never be able to leave the gravitational field surrounding a black hole. As already described, radiation/light is not particles but mediated as waves through the dark matter, the Ocean of Strings that surrounds and penetrates all matter. As the number of atoms and subatomic particles decreases by being crushed, the gravity inside a black hole decreases. This should result in less pressure pushing the strings and that the size is being maintained at a level supporting equilibrium. Should two black holes collide, the combined black holes would subsequently shrink to this size. Without the reduction in size, the whole universe would *already* have been one big black hole. Everywhere, subatomic particles are being created by strings spontaneously joining together. In an eternal process, these particles that are in a close proximity to a black hole are drawn straight in it, emit flashes of light while getting crushed and then become strings again. Compare to "Hawking radiation". **Therefore, as explained here, it is not possible that black holes would have been further compressed to have been part of the so-called primal atom, together with all the galaxes.** Thus, the Big Bang is hereby a proven figment of the imagination. Claiming its existence should therefore not be accepted.

Old Theories

Hence, as described above, the black holes do exist for a logical reason. Black holes pressing out strings would incidentally align with Einstein's 'General Relativity Theory' and its hypothesis. A hypothesis that has given a rise to wild speculations about "**white holes**" that would be connected to black holes through magical "**worm holes**", and opposite to the black holes, they would emit matter. These worm holes have been object to fascination and hopes that these would also function as canals spanning different times. Believing in this and believing in the possibility to somehow affect the passing of time by traveling at a high speed is based on the fact the relativity theories include Lorentz's incorrect interpretation of the experiment conducted in 1887, described in Section 2 'The constant speed of light, and Einstein'. Lorentz's factor is discussed in the Annex.

Nobel Prize 2020

Furthermore, results from mathematical expressions for the above theory indicated incorrectly that a black hole would be immensely small with infinite mass. This was something that not even Einstein himself believed in because the concept "infinite" is physically impossible. In 1965, Roger Penrose succeeded using his **own mathematics** to show that black holes must exist. And in 2019, an astronomist succeeded to show how stars in the center of our own galaxy are revolving around something invisible, possibly a black hole. In the same year, another astronomist was able to photograph the contours of a black hole located in the galaxy Messier 87 with help of his colleagues and a global network of radio telescopes. The following year, all three shared a Nobel Prize they received for their achievements. **Similar to what is claimed in this book, Roger Penrose believes that there never was any "Big Bang".**

Dynamics of the universe – Problem statements

It is easy to realize that a universe in equilibrium must have forces opposite to black holes. The process of spontaneous fusion of strings provides matter, and black holes re-build strings. The above relativity theory describes also "curved space" which refers to the universe as an infinite "∞" with nothing outside. How well does this form fit the Big Bang fantasies? And judging by the latest three-dimensional model of the universe, based on intensive mapping, it does not seem to have that shape either. However, this book has already rejected Lorentz´s factor, and subsequently this theory. The definition for the universe of being everything, without "something" on the outside remains. Integrated in this definition is the statement that energy-mass cannot be lost but it remains constant. In other words, the universe is a "perpetual motion machine" that keeps processes in motion without consuming energy. In addition, observations leading to conclude that the universe is expanding, even accelerating, have been made. But this applies only to the part of the universe that we can observe. The expansion cannot be caused by a repulsive force between stars and galaxies. The explanation could be, as mentioned, that the universe rotates and hurls galaxies outwards, or by the emission of radiation from the stars that causes them to be pushed towards empty regions without dark matter. The consequent effect will thus be accelerating. Another, less likely explanation would be black holes in the universe's "horizon". In this case, to remain in equilibrium, a universe would need an influx of strings (quasars) released from these black holes. And what would possible be behind these black holes? Here, at the outermost edges of the universe, edges that by definition should not exist, may philosophy and populistic explanations take over.

String Storms

It is not completely impossible that storms can blow through the Ocean of Strings caused by differences in pressure originating from black holes among other things. Such storms should exist and have an impact on the gravitational fields in the universe.

Materialization

Upon being pressed out and freed from a black hole **the strings** will spread out in the process of equilibrating pressure in the universe. Initially, the strings merge to form subatomic particles, such as protons and electrones, which can be the very process identified and characterized as **quantum fluctuations** or **vacuum state fluctuations**. Based on the overall comprehension provided by this book, one could call it "**materialization**". Before the protons and electrons find each other, at least the protons form clouds consisting of something that must be characterized as **ionized hydrogen gas**. This has indeed been observed, indicating that the strings in certain areas of the dark matter in the universe are sorted.

An explanation for this sorting could be that where the black holes push out the strings, the building blocks from the crushed subatomic particles, there the density is greater than the strings that are only part of the protons. Electrons could be formed spontaneously from surrounding dark matter, and over time captured by the protons, i.e. the ionized hydrogen, and over time form large clouds of hydrogen. These are then being drawn by gravitation into larger and larger clouds that finally form a star. Relatively recent observations made by Alma, the radio telescope located in Atacama Desert in Chile, confirm that new stars are being formed in the quasars in which the detached strings have been pushed out from a black hole. Based on the movement from the black holes, it is assumed that these stars are being formed at a "safe" distance from the black hole.

The Eternal Process of the Universe

Forming subatomic particles is a natural and **continuous process** resulting in merged strings with a measurable mass. **This process is ongoing not only in close proximity to black holes but also in and arounds us**, and close to our own sun, "feeding" it. Maybe a star will have time to burn out, explode as a supernova, and produce planets before it is seized again by a black hole that squeezes out strings, and by doing so, "closes the circle". These cycles have revolved an infinite number of times, for all eternity. Not 14-15 billion years (a naïve thought) – just because our telescopes do not see past these light years. Solar systems and life are born to then be swallowed by a black hole before they again are being squeezed out as strings of the dark matter, the smallest building blocks of the matter.

Figure 15. A **quasar**. The illustration is based on data obtained from ESA.

Figure 16. When the pressure form a so-called thermonuclear reaction suddenly overcomes the gravity of a star (i.e., colliding stars), an explosion, supernova, is the natural consequence. Due to the high temperatures, it is believed that the heavier elements are created here. A picture from NASA.

The riddles of physics lead to assumptions, religion and philosophy.

In the forever silent universe in a deep sleep – a seed of awareness began to grow.
Awareness that, in the depths of the dark matter and through the endless time, grew,
became infinitely vast encompassing everything.
And thus, it was just one, and it was lonely.

Of the dark matter, matter begin to form,
permeated by the infinite awareness.
Life was born. Life developed.

Life was given a voice,
and the awareness could no longer be heard.

About consciousness.

This book has shown that we are surrounded by dark matter, the strings/chains of strings, which also are part of the structure of atoms and molecules. Also; Under the heading "Parallel worlds" the so-called supersymmetry is described, and the possibility that some groups of dark matter strings form particles, which do not bind to the particles that make up our world. Could such a parallel world, a "shadow world", affect us when their particles and strings interfere with "our" atoms?

Electric current influences dark matter to build magnetic fields. Correspondingly; changes in magnetic fields create electric current. Matter and dark matter thus interact. Activities at the biochemical level imply activities at the atomic level, which is reflected in the dark matter connected to the atoms. All biological life thus affects associated dark matter. With the interactions described, it should mean that dark matter can control biological life.

After all, we are controlled by what happens at the cellular level, which is controlled by what happens at the molecular level, which is controlled by what happens at the atomic level, which is affected by the aforementioned interaction between atoms and the strings of dark matter. One should therefore be able to assume that we are not isolated from an external and invisible influence. If you e.g. decides to raise an arm, you cannot always, at a so-called impulsive action, be sure what is the underlying cause of that action. Events that are connected due to cause and effect through time and space, we can explain. But could an impulsive action instead be explained by the influence described above?

Where is our consciousness? According to **Hans Zingmark**, chief physician at Halland Hospital in Halmstad, **no part of the brain has been able to be determined as the home of consciousness**. Is consciousness located in the dark matter that is connected to our biological processes? Zingmark, in addition to his daily work as a doctor, has also conducted a study of near-death experiences, in which he has interviewed patients who have been revived after all electrical activity in their brains has ceased. In that study and in the coincidental accounts of others, experiences such as being out of body, seeing one's own “dead” body, encounters with deceased relatives and friends. Read more about referenced studies, for example on the internet!

Is consciousness a copy of every living body according to the said theory of supersymmetry? Is consciousness our "soul" that survives the body? What constitutes and harbors the God that man has instinctively sought in all ages? Can experienced local temperature drops at so-called paranormal experiences be explained by reduced kinetic energy in the strings of dark matter? Within the so-called science relevant questions should be asked, and the assessment of what is relevant must never be clouded by prejudices and ingrained beliefs. The latter is a problem with authoritarian-controlled research, and an explanation for why paradigm shifts in science require personal initiative.

Regardless of the answer to the above questions, we are made up of strings of dark matter, which are influenced by other strings in the ocean of strings, which is our home and building material both for our bodies and probably also for the shadows of our bodies, our consciousness.

I am I, and you are you,
a wave told another,
before it hit the shore.

And like small ripples,
they were taken up by other waves in the everlasting ocean.

Infinity.

Although there are believed to be 17 billion Earth-like planets in our galaxy alone, many religions have designated a creator and ruler of the entire universe. The existence of a more locally, in time and space, anchored spiritual ruler - since this creator would have man as his main task - is for many religions, and in accordance with the nature of man, not ambitious enough. However, this would not be a logical problem, if God can be considered the all-pervading consciousness.

Under the heading "The Big Bang can never have happened", the concepts of infinity and eternity were touched upon, concepts that are difficult to grasp. Especially when the time man has existed is not measurable in the history of the universe, and the planet earth is not visible in the infinite universe.

When, after biting into the sour apple of knowledge, one realizes the logic and correctness of the fact that both time and space are without limits, it often leads to a feeling of the nothingness of man. One would like to believe that man's place in history is greater, e.g. by proclaiming that our world is not that old, and that the earth is the center of the universe. But it may be helpful, with the apple in your mouth, to mentally try to embrace the understanding of the ocean of strings, mother of all universes, and adopt a philosophy which consists of enjoying life as much as possible, also outside the credulous paradise. Because the risk with knowledge is that it comes at the expense of blissful ignorance. If that doesn't help, there is always the option of choosing a religion, that makes the individual feel as being an important part of a chosen “universe”.

And the philosophy.

Suppose that in “our” part of the infinite universe the ocean of strings lies with a diameter of a few tens of billions of light years or more, and that in this ocean all the galaxies lie. Suppose further that a traveler makes his way past all the stars and finally also past the outer edge of this ocean of strings.

When the traveler finally can no longer see a single star and continues to travel, does he get anywhere or nowhere?

This getting nowhere ought to be, philosophically speaking, the outer limit of the universe. But there is nothing to prevent the existence of several separate ocean of strings, i.e. dark matter. The interesting thing about the possible existence of separate string oceans is that neither light nor gravity would be possible to transmit between them. It should be recalled that; Strings, as this book has shown, mediate both light and gravity, but are not themselves subject to gravity, except in the form of congestion in black holes, from which they are squeezed out. The strings that would possibly detach from the outer edge of the ocean of strings, would not be pulled back. This points to the fact that after an eternity there would only be one single ocean of string with reasonably evenly distributed density. True, the universe could be infinitely large, but does it contain infinite amounts of strings?

It has been established that the universe's galaxies are clustered in different groups, so-called **galaxy clusters**. One could imagine that each galaxy cluster corresponds to a separate ocean of strings. The fact that we can still see other galaxy clusters could, in that case, be explained by the fact that there are also strings between said oceans of strings, but with a lower density and thus a lower speed of light. Alternatively, this space could consist of other things, for example strings that do not include "our" strings. Also in that case, there would be a different speed of light. Intergalactic distances, which are measured in light years, take on a different meaning in both cases. At the same time, there may be oceans of strings with stars/galaxies that are so isolated that they are not surrounded by anything that could transmit any kind of radiation, including light. **Such "hidden" ocean of strings also conceal possible stars in the background.**

Regarding philosophy: The possibility that there is a copy of Earth with a copy of us in another solar system is infinitesimally small. Mathematically, infinitesimal is described as $1/\infty$, i.e. 1 divided by an infinitely large number. But now there also is an infinite number of stars (suns) with planets in the universe. We therefore have to multiply $1/\infty$ by ∞ which then gives the formula: $1/\infty \times \infty = 1$. So the possibility that there is a copy of us on a copy of Earth is 100%.

Figure 14. A spaceship leaving the ocean of strings (dark matter) with galaxy clusters behind.

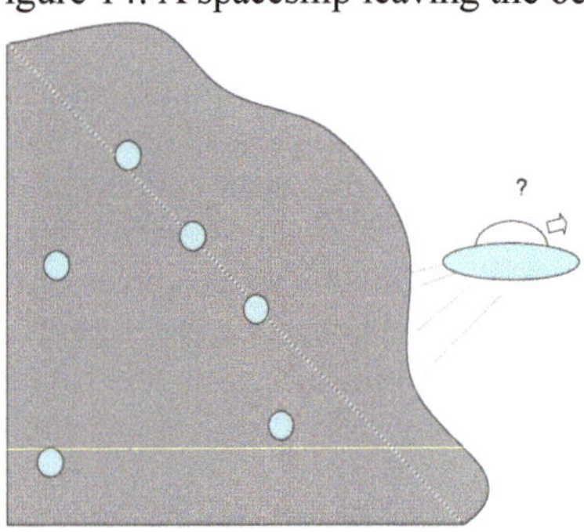

In an intergalactic space with a total lack of strings (dark matter), the speed of the spaceship will not be limited to the speed of light, but then there is nothing for the space ship to orient itself by, to measure its speed against. Or are the oceans of strings housed in an even larger ocean consisting of an even more finely divided material of which the strings would possibly be made up?

Postscript

The contents of this book that have now been reviewed including the conclusions that have emerged are revolutionary, and likely also provocative, for anyone who has been caught in the conventional thinking. The core of the book comprises a scientific milestone, and because of its magnitude, it will not become accepted automatically by scientist working on the field. Some of them might lack the curiosity and investigating attitude characterizing a scientist. Furthermore, independent thinking inside an establishment is associated with risks. Like a mole, who does not need an overview of the landscape, digging through the old tunnels established by commissioners, they take the safest road. (The author has the “unfair” advantage of not risking anything with the clear description of the landscape with this book.) History has shown that any large fundamental changes in concepts need time to let the shock settle and leave space for common sense and acceptance. The problems Charles Darwin and Galileo Galilei faced are examples of this.

Without directly violating all established concepts about an atom’s internal physics, the book’s conclusions provide a fresh awakening to a new day open for new possibilities. We have finally received the long-awaited explanations to the nature’s innermost essence – at atomic level.

Summary of the book

The story could have begun by pronouncing the photon dead, resulting in the following logical string:

1. The photon does not exist.

2. We have established an “Ocean of Strings” that mediates wave motion.

3. String movements will hence become responsible for the inter-atomic interactions and impacts on the atomic level, which has previously been explained by existence of photons.

4. This interaction and influence shows therefore that the strings can affect what happens inside and between atoms. To explain this influence by strings “tackling” each other only would not cover everything that happens in an atom, or enlighten what controls this. But since the photon does not exists and hence it cannot be used to explain intra-atomic events, the strings have been assigned a role where they can latch on the subatomic particles “string handles”. This role seems very natural if the named particles are built up of strings (which many people have accepted). String handles sticking out from something that is constructed of strings seems equally natural.

5. Once the concept “string handle” was established, all pieces fell in place, and now it became possible to explain all four natural elements including the mechanisms that are behind electrical current. An additional bonus was obtained from adopting the bold assumption about inter-electronic string handles that makes it possible to explain what controls the chemical processes at the very basic level, and to provide a plausible explanation to why atoms with 8 electrons in their outermost shell are stable. It is this explanation depicting the stable cube form that confirms the theoretical construction so well that the theory can be sealed with the seal of credence.

With the string handles and gravity explained, the photon as the explanation for weight difference measured upon fusion/fission of atoms became redundant. The logic can hence be stitched together to make it impermeable. Pieces of the puzzle find their places, and the most interesting piece is the cube because it describes so simply and so accurately what is driving chemical processes.

We have finally in our hands the needed, cohesive and holistic theory that explains all four natural elements. And it works!

Because, what it is *all* about, and what the four natural elements all boil down to is this;
moving strings that latch to, or do not latch to, each other.

And whether this is controlled solely by coincidences generated by the chaotic noise in the Ocean of Strings is a question for religion or philosophy to deal with.

While analyzing kinetic energy, we developed a formula that describes its relation to velocity better. A formula, that takes into consideration the strings surrounding matter, and that simultaneously provides that same result than the conventional but less logical formula.

After reading this book, you might also note that the theory of a "Steady State" universe and the theory of the light-mediating "ether" must yield for an incorrect interpretation of the reality. The role of the strings and their relation to matter has never been understood by those studying the field. They have even had difficulty accepting the atom. It in only a little over one hundred years since Boltzmann was met with shameful arrogance when he tried to launch his idea of existence of the (physical) atom. Boltzmann took his life a few years later. The "authoritarian thinkers" are of course interested in maintaining the worldview as it is and not letting other views through. Therefore, it should not be a paradox if salary-lifting scientists often wear blinkers. **However, there will be a day when we all wonder how the scientists could believe in a variable speed of time.**

I finish by reminding about Thor Heyerdahl's Kon-Tiki expedition. The self-righteous so-called scientists refused to accept his evidence showing that Polynesia was populated from South America, despite the complete lack of evidence supporting immigration from Asia. When Kon-Tiki expedition finally reached Polynesia, Heyerdahl must have been overwhelmingly emotional.

And after forming the theories described in this book, I came to the following conclusion: If He_2, contrary to expectations, exists, the molecule must show signs of weightlessness. The following discoveries supporting the assumption came to be my emotional "landing in Polynesia". The discoveries demonstrated not only that He2 exists (although only short moments in laboratory) but it shows all the signs of zero gravity.

Rolf Sjöström.

Question authority!
Thor Heyerdahl

Annex

Additions, which touch and belong to the book.

Lorentz factor; a figment that has distorted our worldview.

This speculative hypothesis was born from an experiment conducted at the time of "kerosene lamps" in 1887, in other words, during the relatively "uninformed" time. The fact that the hypothesis is a scientific speculation that would not be accepted today stays hidden for the most under a "layer of dust" collected over decades plus one layer of dazzling mathematics. Thus, the factor is not being looked at by critical eyes, and has over the time become part of the large and well-established theory. So established that its status has been elevated to be the truth. The unfortunate consequence has been generations of thoroughly brain-washed scientists of whom not even the most alert individuals dare to question the hypothesis, and therefore, they will stop reading right here. Hypothesis was formulated as the following formula or factor named after its creator.

$$\gamma = \frac{1}{\sqrt{1-(\frac{v}{c})^2}}$$

The above formula shows that the factor's (assigned a Greek alphabet) value for an object changes depending on its velocity 'v', where 'c' is the speed of light. In complete standstill, the value is 1 and increases along with increased velocity. At a velocity, close to the speed of light, the factor's value becomes an infinitely large number. Depending on what you want to calculate you can either multiply or divide with this factor. For example, $m=m_0*y$ is thought to show how the mass would increase along with increasing velocity. Despite the extensive use of Greek alphabets and an alibi provided by mathematics describing "Riemannian geometry", or associated "tensor", further developed by "Ricci" – the misinterpretation of the 1887 experiments cannot be justified. **The mathematics is correct but Lorentz used it to describe his misinterpretation. This is <u>not</u> a way to correct <u>any</u> incorrect interpretations, and this should be self-evident both kids and grown-up scientists. But this was, hopefully unintentionally, left hidden behind the mathematical smokescreens.** It is called magic when magicians use this kind of methods to capture the attention of their audience. The magic with mathematics performed by the involved gentlemen was rather novel, indeed. And since both Riemann and Ricci were Lorentz's contemporaries, a lot of novel things were presented simultaneously and that could have caused some confusion among the scientific elite of that time. Instead of trying to solve the underlying problem with the mathematical formulas, they have been "superficially polished".

Lorentz factor was established as one of the cornerstones for Einstein's nonsense about how the time for an object is affected by the object's velocity. We are now able to convert this cornerstone into a wedge to topple pedestals of nonsense. Here follows the reason why this shameful factor was created and why we are now able to replace it with a correct worldview.

This factor was created by Lorentz in the end of 1890s to explain the results of Michelson's and Morley's experiment from 1887. The experiment aimed to demonstrate with help of mirrors that light travels faster in the direction the Earth orbits the sun, and slower in the opposite direction. Because light was thought of as waves moving through the ether (what we today call for "strings" or "dark matter") the experiment was supposed to reveal that the ether was connected to matter, in this case to the Earth. The thinking before the experiment was that a light wave was incorporated with the Earth to one and the same coordinate system in which the sun was in the "origo". If the ether is not attached to this matter, the light would come in at an angle moving back to reach the "back" mirror first. The light reached both mirrors at a same time, hence, the ether must be attached to the matter. For Lorentz, this was not that simple. In his opinion the ether should be thicker closer to the Earth's surface because of the gravitational force, and as such, it should affect the velocity the light is traveling. But since the speed of light has already been determined to be constant and seemingly unaffected close to the Earth, he thought that a different explanation was necessary. One would be that maybe light does not consist of waves traveling through the ether,

Lorentz has now commit the unforgiving mistake to take for granted that the ether is connected to the Earth <u>through</u> the gravitational forces. However, this had not been established.

This was the beginning of the scientific derailment. The chapter 'Role and characteristics of the strings' of this book describes and summarizes the connection between matter and the so-called dark matter. The derailed train was pulled further over logs and boulders with a help of fancy and impressing mathematics while hiding the derailment. The great question is though why is an outsider needed to point out that the Emperor is naked. Maybe because those busy with daily grind as a consequence of the derailment would face reprisals if they questioned the Emperor's cloths. It could even be considered a betrayal of the workplace "woven" in the web of Lorentz factor.

While searching for an explanation, Lorentz came to think about Newton's logical formulas to calculate objects' positions with placements that are described in separate coordinate systems. Newton had named these formulas "Galilei transformations" after the Italian astronomers and "transformation" because coordinates from one system are translated (transformed) to another. It became easier to calculate a planet's position in relation for example to the Earth's coordinate system.

Lorentz realized that these formulas were relevant for interpretation of the experiment. Since the experimental setup was traveling at the Earth's velocity as the Earth moved in its orbit around the sun, the result must be impacted by this velocity. This was logical and expected if the ether did not follow the Earth's along its orbit. But Lorentz did not stop here; he took a step out to the speculative darkness by making the following argument to explain the result obtained from the referred experiment: "Speed of time for an object is affected by the object's velocity". To describe this, he adopted the mathematics mentioned in the introduction.

Knowing all this, it is uppermost strange and surprising that Lorentz factor is still uncritically accepted and allowed to pass through the needle's eye usually firmly held by critical scientists who have the power to decide over the truth.

The greatest damage was done by the factor when a major opportunist inserted it in mathematical descriptions of the reality. This led to spectacular and crowd-pleasing conclusions, the relativity theories – "Emperor's new clothes" – which even the mass media is afraid to question.

An interesting phenomenon is that a construction made of theories can overshadow an error in the one theory all the other theories are based on.

Power of nonsense and the arrogance fall when the fool's stone is grabbed away, and the footmen lose their job.

Logical analysis of Lorentz factor results in unsustainable conclusions

One could now ask whether more than one speed for time is possible within one and the same system. As an example, if we look at our solar system as an inclusive and superordinate system with one timekeeping, and at the same time we believe that this system contains one or more parts with a different timekeeping, this would mean that a singular object within the solar system has **simultaneously** both an individual speed for its time, and another velocity for the time that applies to the object when seen as part of the solar system, the superordinate system!

In addition, Lorentz's reasoning on measurements of the velocity leaves many unanswered questions. According to logic, speed is unquestionable and always relative and therefore it must be measured in relation to another object. If Lorentz factor is applied to two objects moving away from each other at high speed, then Lorentz factor as an explanation becomes scientifically unsatisfactory. Maybe one of the objects is not moving at all? Further, for which of the two objects is time traveling faster? And how can one know that acceleration in the space for an object is not retardation? Those who attempt to explain this using mathematics based on Lorentz factor have not understood the problem, but the figment has instead been explained with the same figment!

If something other than light beaming towards the Earth was chosen for the 1887 experiment to test velocity, would the velocity of this something else have become the magical velocity 'c' which Lorentz factor is based on. We could run an experiment using shock waves through water in a pool where the waves travel perpendicular to the Earth's rotational direction. In this thought experiment, the shock waves would have been initiated by a rocket. The rocket would travel at 30 km/sec against the Earth's rotational direction and fire every time it passes the pool. Impossible to carry out for practical reasons, the alternative to the firing rocket could be a smaller "firecracker" moving closer, and using its own coordinate system. Next, I constructed "Sjöström factor" (named after the author) that can be copied into the above described factor in, in which 'c' is replaced with 'a' (aqua (finer language)) for the velocity of a shock wave moving though water. The firecracker could create a shock wave in a plastic tube directed towards the middle of the pool. And like 1887, two different densities are dealt with this time too, air and water, through which the shock wave needs to pass. This applied Sjöström factor would make the time to move very fast, already at a lower speed. This thought experiment demonstrates how ludicrous Lorentz factor really is!

Lorentz's argument was indeed not accepter just like that. It was subject to discussions spanning three decades but because no other viable explanation to the referred experiment was available at that time, the exhausted scientists were reluctantly forced to accept the argument which raised more questions than provided answers. Any scientist with a carrier kick-off after this turbulent time was disciplined to accept the agreement as the ultimate truth, and the criticism was silenced. It is not difficult to understand how religions are born, churches built, filled with high priests to manage the convention for the truth and to contest other ideas, especially those that could topple their church. To contest even ideas that would reflect the truth initially sought in 1890s.

Einstein's Two Postulates.

Einstein based his relativity theories on two postulates. One of them states that all reference systems and their coordinates are equal. The universe and its ether (dark matter) were thus deprived their role as the absolute reference system to which all other reference systems relate. The other states that the speed of light is constant.

Background.

The postulates and theories were published after scientists, such as Lorentz, were unsuccessful trying for 18 years to explain the results obtained from Michelson's and Morley's 1887 experiment. See 'The constant speed of light, and Einstein' in Section 2. See also 'Lorentz's factor; a figment..' in this section. **The first postulate** states that all reference systems are equal, in the sense that, for example, outer space and the equipment in said experiment have, to each other, independent system coordinates. The extension of that reasoning means that there can be no ether with a common parameter for time connecting these systems, and **that the ether therefore cannot be the medium light is using to propagate its motion.** Subsequently, the assumption that light consists of particles must be adopted. As a result of this assumption, it became impossible to explain how these particles can maintain a constant speed regardless the speed of light source (and its "system"). This imposed the second postulate to describe **the speed of light is constant**.

The researchers had previously assumed both that the speed of light is constant and that the ether is its medium. However, they encountered problems accepting this, when they realized that gravitation compresses ether, which unfortunately had been taken for granted. But if they had understood the true nature of ether (the strings), as explained in this book, we could have avoided both Lorentz's factor, Einstein's postulates and any related theories. See "Role and characteristics of the strings" in the end of Section 1, which concludes with: As the strings are being bound to adjacent matter, in a way that they only hold matter together, they mediate gravity, and thus do not compress itself. And the property of being bound to matter, as for example air, also explains the result from the referred experiment. **Important: Einstein did not find a solution! He created one!** And describing these fantasies with mathematical formulas does not turn them into truths.

Acceptance of an explanation just because it cannot be counter-indicated.

The postulates belong together because both are needed to explain the said experiment. This is confusing, and to solve a problem with a postulate should be deemed unscientific and thus Einstein should have been expelled from the scientific arena for life. The relativity theories are based on these two worthless postulates and on the Lorentz's factor. The theories were naturally met with resistance by contemporary physicists. The reason why the theories nevertheless got hold might be the fact that nobody was able to explain the result of Michelson's and Morley's experiment. Yet, it is surprising that modern-day physicists can accept an explanation as a truth just because they cannot prove the opposite. This seems to be customary. Similarly, for some time ago, it was not possible to prove that the Norwegian god Thor was not behind the thunder and lightning.

Mass Media vs. Serious Critics

The critics against Einstein's theories kept up their position for a couple of decades. But he had his younger age to his advantage and survived his critics. It might have also been that his personality helped to market him better than his critics in the mass media. Even so, articles containing exciting fictional theories have had a higher commercial value than comments by critics. Articles written by journalists with varying knowledge levels are accepted as true by the public. This gave rise to the possibility that a convention could be shaped by mass media's economic interests. The young students who would carry on the knowledge in physics of that time were forced to learn the new teachings. It was equally risky then and today to question theories established by someone famous within the field. A few generations of particle physics later, the questionable and far-fetched theories had unfortunately turned to solid unquestionable "facts".

New Knowledge – But No Courage To Re-evaluate the Postulates

Despite access to new knowledge, scientists are not brave enough to re-assess the 1887 experiment with open mind. They face the risk of concluding that both postulates can be condemned and the Lorentz's factor ends up getting trashed. Observations showing that the dark matter both exists and is bound to matter seem mount to nothing. See Figure 11a and Figure 11b (pictures) in Section 4. Moreover, quasars indicate that the dark matter cannot be compressed opposite to what Lorentz believed. See "Big Bang Could Never Have Been" in Section 5. To be or not to be – that is the question about the dark matter, and most certainly still subject to discussion along with its character among scientists. But what is a breach of logics is a postulate that does not accept the definition on an all-embracing universe or the portion that we have oversight of, such as an overarching and absolute reference system to which all other reference systems included therein can be related. An explanation should not be accepted just because it cannot be counter-indicated. A later mathematical description to this explanation will not make it to represent the reality.

The above logics is unfortunately ignored arrogantly by modern-day particle physicists. They consider themselves smarter than Bohr, Stokes, Planck, and other scientists who did not agree with the imaginative Einstein's interpretation of the referred experiment.

Would particle physicists approve the postulates today if these were presented to them for the first time? Most likely not. Would they understand and approve the description for the **"Role and characteristics of the strings" provided in Section 1 of this book as more logical than the said postulate**? Again, most likely not, and this for many reasons. First, the scientists in the field have well-developed blinders. Second, their career selection is based on interest, not on intelligence. Keen interest in one's profession regardless what they do is a great benefit and can sometimes compensate any shortcomings in intellectual capacity. But the blinders remain, and they should be the last thing we wish to see being worn by those we pay to make discoveries. Blinders comprise an institutional weakness and are to some extent forced upon individual scientists. Read more about the topic in "Intelligence – Knowledge (IK)..." in this section of the book.

The reason why the more than a century-old postulates still remain accepted is that they are handed down as established "truths", and that no one dares to risk their employment by telling it as it actually is – that the Emperor has no clothes.

Today, we are better prepared to explain the 1887 experiment.

Today, we know that the atom exists. And the referred book shows how the ether (strings/dark matter) is attached to the adjacent matter (atoms). The book demonstrates also that these strings, despite their attachment, are not affected by gravity, and therefore its density is not affected by the distance to the Earth, either.

The book, with its logic based on results from well-known experiments and with the knowledge of the stable atom with 8 electrons on its outer shell, and other established facts, guides us to understand a more modern explanation. The logic is constructed like a rock climber's anchoring points in a mountain. With the help of these points anchored into the current science, the book leads us to the following conclusions:

Role and characteristics of the strings

1. Building blocks for all matter are composed of strings. Hence, the mass is derived from all included building blocks, and these interact with the surrounding strings. (Get latched or get not latched.)
2. Despite their mass, the strings outside matter **are not affected by the gravitational force** but they create and mediate gravity between matter – *like a chain shaking between two knots (matter)*
3. Therefore, the string density will not increase by gravity because the strings themselves are the mediators of the force we experience as gravity.

Conclusions on the above items

4. Strings are coupled to all matter; not because of gravitational force but with the help of hooks.
5. And this explains the result Michelson and Morley obtained from their 1887 experiment that tested whether the ether follows along the Earth or not.
6. This explanation eliminates the need for the figment, i.e., the variable velocity for time.

In his pursuit of evidence supporting the book's theories, the author decided to create a virtual zero gravity molecule. The illustration that begin to take form had 2 helium atoms linked together. See the image of the molecule in the book. Upon researching online, a hit was found of this molecule that was, indeed, weightless. The molecule arises upon cooling helium down to close to absolute zero. See
http://www.youtube.com/watch?v=2Z6UJbwxBZI
Observe the expression you hear in the video: "It appears to defy gravity!"

Occasionally, here and there, you hear "Hallelujah" for example for gravitational waves, and how the so-called atomic clocks in GPS satellites run slower because of their velocity in their orbit around the Earth. This is nothing but new nonsense based on existing nonsense. The book explains gravitational waves. And about the above clocks, they are inaccurate enough to require regular adjustments. If nothing else, these clocks can be affected by the flow-through of strings, penetrating through the string cloud already attached to the satellite. (Described in the book.) And the equations that apply to the so-called atoms, apply a for waves moving through the 'Ocean of Strings', which the visible universe is part of. This too is shown in the book which includes references to recognized research. By means of the conclusions drawn up, all the pieces of the puzzle are finally falling in place.

Theses for ‘Role and characteristics of the strings’ nailed – 500 years after Martin Luther nailed his theses on the door.

Summary of how the book relates to history.

Historical mileposts ← Ether = Strings → A new era

1818 → According to a theory Jean Fresnel launched, light consists of waves through something he called for "ether", and the ether penetrates everything.

1864 → Maxwell constructed a formula to calculate the speed of light, the waves moving through the ether.

1887 → To answer the question whether the ether was partially or fully attached to the surrounding matter – discussed since the beginning of 1800s – Michaelson and Morley performed an experiment in 1887 where mirrors controlled light from one source so that one beam traveled in the direction the Earth orbits the sun, and the other in the opposite direction. Any interference would demonstrate a difference in time needed. There was no difference, and they concluded that the ether moves with the earth. The result was consistent also with Stokes' theories.

However, Lorentz could not accept the interpretation. It did not agree with the assumptions and calculations on the ether's relation to gravity.

It was here where the development got derailed. In lack of better ideas, and based on Newton's Galilei transformations, Lorentz began to build theories about a variable velocity for time.
Discussions had lasted for a lifetime, and the "old men" were worn-out. Therefore, not all of them had the strength to protest when young Einstein picked up the baton after Lorentz, and transformed Max Planck's concept, light quantum, to a new particle, the photon. The rest is well-known history with its challenges that, interestingly enough, have been taken for granted.

This "derailed train car" can now be cleaned out of the way to let the development move ahead.

Traveling back in time revealed that Stokes got it most right. On the contrary, everybody got the theory of relationship between ether and gravity wrong, which led researchers in the cul-de-sac that became their base. Despite big difficulties, they have succeeded to find "small islets of truth" but without knowing how to read the map.
The theory presented in the book for the role and characteristics of the strings is based on the atomic structure. **To defend the above men, the following needs to be mentioned: the configuration of an atom was not known at that time.**

Now → **A possibility for a new start for research on these fields.**

2011 → This book develops the theories of strings and their interaction with matter.
The coherent and uniform theory that explains all four natural elements is found. At the same time, the book develops theories that, among others, doom the photon redundant.
Many experiments conducted by established institutions both prove, and are explained by, the above theories. An example: The result of the 1930s experiment in which helium was cooled to helium II, can only be explained by the theories presented in this book. **See Section 3!**
Without prior knowledge about theories from the 1800s about light traveling as waves through the ether and about interaction between the ether and matter, the author came to draw identical conclusions by relying on logical thinking. But this time, the historical mistake will not be repeated because:

In parallel with the coherent and uniform theory, the author developed also a sound theory about the role and characteristics of the strings.

With the historical derailment behind, along with awareness of the strings, the scientific development can go forward.

Therefore, the book's theories can explain effects previously explained by theories of slowing of time.

In addition to this historical intervention, the book demonstrates, along with an associated formula, how the strings, so-called dark matter, interact with matter in conjunction with generating kinetic energy. The formula is based on the logic that velocity is always relative, and shows how much mass of the dark matter gets stuck on and is dragged along a certain mass of matter at a certain velocity.

The supporting theory –

The book builds on strings and their property to latch on each other resulting in longer chains and compounds in form of different subatomic particles. This is nicely illustrated by a dusty and sunny room where the dust hanging in the air represents the strings in the universe, in the 'Ocean of Strings'. Another example on spontaneous formation of chains and aggregations is illustrated by the below image of pollen.

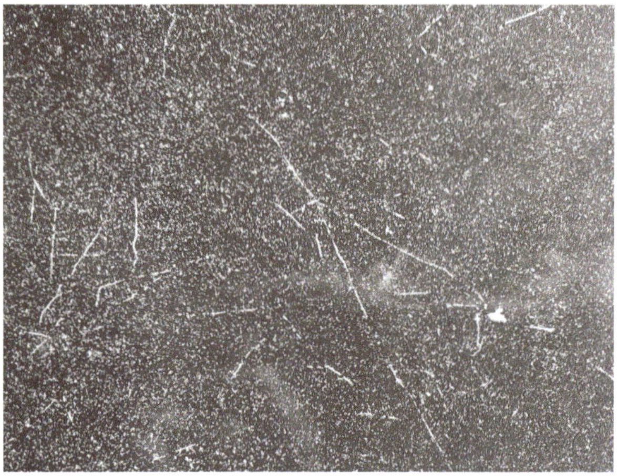

towards the goal.

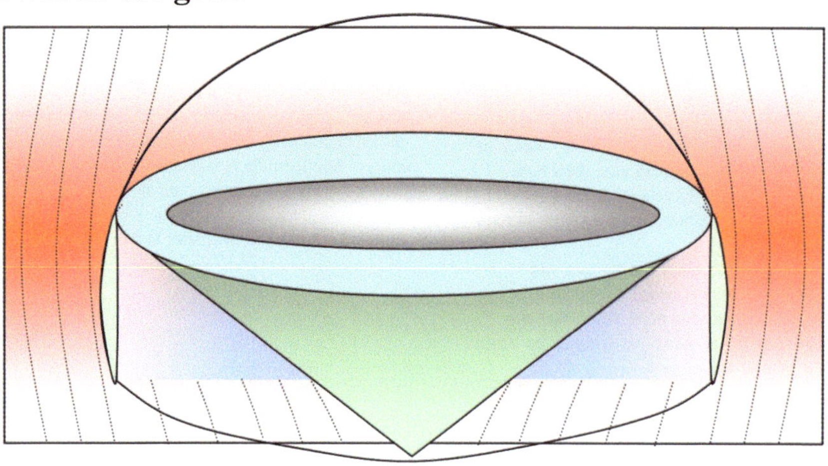

The dilemma – authoritarian arrogance towards the common sense.

The flipside of arrogance – foolishness

It was the authoritarian dictatorship within (non)science that tried to stop Galileo Galilei when he argued that the Erath was not the center of the universe.

It was the authoritarian dictatorship within (non)science that tried to stop Wallace and Darwin's ideas about evolution of plant and animal species.

It was the authoritarian dictatorship within science that tried to stop Ludwig Boltzmann when he presented his (and John Dalton's) theory about the atomic structure.

It was the authoritarian dictatorship within science that tried to stop Alfred Wegener when he presented his theory about moving tectonic plates.

It was the authoritarian dictatorship within science that tried to stop Thor Heyerdahl when he presented his theory about the origins of the Polynesian population.

To understand the above examples (among many others), we have to respect the fact that every man is "driven" by his DNA which for the most part is identical to his hairier cousins' DNA. Whoever is on top at a given moment must keep his group together and guard his status. What might be best for the group is determined by self-interest. Subsequently, any scientific argument counter-arguing a mainstream theory can never count on getting an objective evaluation because those guarding the status quo will be distracted by the foreboding threat closing in. The paradox: "one with hunch judged by dumb" is hence nothing less natural than that all wars can be traced back to the primitive streak embedded in our genes. This dark side of reality has claimed many innocent victims among the truth-seeking heroes. While it is not a happy story, it is surprisingly comical once you realize the following:

Even though any confidence the authorities have enjoyed should be gone by now, are the masses with their need to belong to a group as believing as during the dark middle age. And instead of accepting the simple and convincing explanations provided by this book, many prefer (also for reasons of convenience) the accepted fairytales, such as:

- A so-called photon is a particle which also knows what another photon will do!
- All the galaxies in the universe once fit inside a so-called primal atom! (How can black holes be compressed?!)
- The speed of light is constant, also in relation to an object that for example follows the light at half of its speed, even though no velocity can be constant if the time concept is maneuvered!

Comically put, but with a more serious undertone: The deep-rooted need to worship a religion, led by the high priests in their white coats, makes the parishioners, us, the gullible, gawking, obedient and devoted, to uncritically believe in any magical "truths". Religious expressions are beyond us to touch. But upon hearing expressions, such as this by quantum physicist Richard Feynman: "it is impossible to understand quantum physics", it is time to wake up, pull the darkness aside and follow Thor Heyerdahl's call to ***"question authority!"***

Ptolemy-Copernicus. Who owns the truth?
What can we learn from history?

Almost two thousand years ago, the Greek Ptolemy, using mathematical functions, succeeded in describing the orbits of the planets and the sun around the earth. This had been a difficult task because the theory does not correspond to reality. But by using a function in trigonometry, he could describe how the planets moved in circles, around themselves, in so-called epicircles, on their way around the earth.

One reason I mention this is to show: That even if you can describe something mathematically, what you describe will not be more true. Compare this with the Lorentz′ factor and the theories of relativity. It is believed that assumptions are true only because they can be described mathematically, and Ptolemy's teachings were believed by "the scholars" until the 17th century. But as early as 1543, the Polish astronomer Copernicus published a simpler explanation of the planets' orbits, namely that it is the planets, the earth included, that move around the sun.

Moral 1: When complicated mathematics must be used to describe something, it may be because you are trying to describe something that does not correspond to reality.

However, no one was interested in hearing anything that did not agree with what they had learned, despite the fact that Copernicus' model eliminated Ptolemy's complicated mathematical descriptions of planetary orbits.

Moral 2: The most important reason why stewards of old knowledge do not want to let through new knowledge, is that it makes the stewards as useless as the old knowledge.

The theories of relativity are based on postulates, which also are nothing but assumptions. Despite that, these assumptions are considered to be a sufficiently solid foundation on which one can build a science. **And that is an inconceivable huge scandal!**

Copernicus' explanation of the movements of the planets in his book gave no cheers. In 1590, the book was discovered by the Italian astronomer Galileo. His own research confirmed the conclusions Copernicus had reached, and he now took the lead in the struggle against the clergy and stewards of knowledge of that time. But even though the clergy, who considered themselves to know best, both threatened Galileo with burning at the stake, and eventually gave him house arrest for the rest of his life, the truth prevailed in the end. Fighting for the truth should not be an ungrateful task. But why did the truth prevail? Well, because in the end so many had been able to verify the truth of Galileo's claim that the priesthood had to give up.

Moral 3: It is not always the wisest who decide what is true.

This book was first published in 2011. Should the contents of this book also wait for almost a hundred years before our high priests accept the facts? That should not be necessary, because you can, using the book's references and facts – and just like Galileo's contemporaries could - see what is truth, and also come to the conclusion that those who rule over the truth also keep it as their prisoner.

Human genes have not changed over the years after Copernicus. This means that the stewards of knowledge, as human beings, are still governed by a vain professional pride and a fear of losing their status. A fear that also contributes to the fact that they do not dare to remove their blinders, but just stick to the established "truth". And here we have the core of the problem. A researcher with blinders is not a researcher.

Can a science based on over 100 years old assumptions and theories be questioned or revised? Is it possible to lift a derailed science, of particle physics in a dead end, up on a scientifically sustainable platform?

Intelligence – Knowledge (IK) – Judgment – What the Group Thinks

Is there a future for present-day particle physicists despite their realm being framed by theories from the nineteenth century?

More than in any other discipline, particle physics relies on interpreting experimental data. A good interpretation requires a good IK in every respect, and occasionally given underlying theories must also be questioned. Any interpretation will always be assessed by human judgment. Meanwhile a person's judgment is under influence by current opinions and theories accepted by the group where he is active in. And because also particle physicists are pack animals, the argument is of interest to this book. An argument that is of general interest, and for some of the readers it might be the most important part of the book.

Figure, which illustrates man's relationship to facts.

Judgment is a product of both intelligence and knowledge, but can be perceived as a separate property, as it can be influenced. High intelligence and a lot of knowledge are no guarantee of a good judgment. Interpretations or deeds can, as is well known, be performed against so-called "Better knowledge".

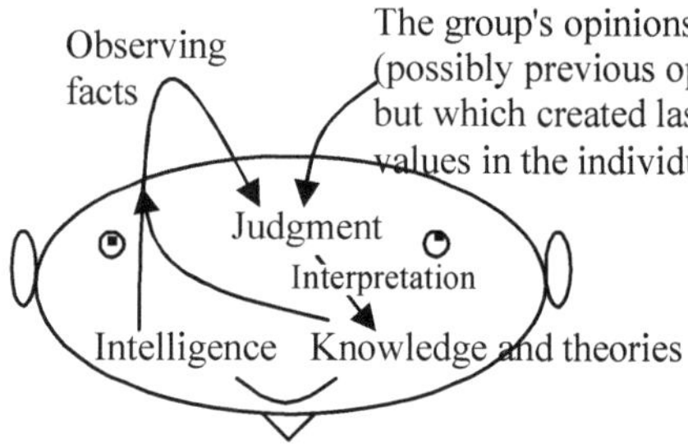

Judgment overrides intelligence

We all have marveled over how, through the history, people with seemingly normal IK have found themselves actively taking part in mobs sacrificing babies during religious rituals, burning women at stakes accused of being witches, and in carrying out genocides. How else to explain this if not with people's judgment?

Homo sapiens, the pack animal

In these instances, the individual judgement has been corrupted by something. To understand what "something" might be we must realize that a human being is a pack animal with a natural instinct to sacrifice their independent thinking not to be excluded from the pack and subsequently lose their chance to propagate their genes. Therefore, high intelligence can be sacrificed on the Altar of Pack Affiliation. This behavior has been evolutionally perfected at the genetic level from canines to homo sapiens. An individual can have the best of ideas on for example a hunt, but it is not worth anything if the pack follows another plan.

Price for safety within a group

Their will to belong to a group can be driven by an instinctive fear to be alone, possibly in opposition to the rest. The benefits of belonging to a group seem to outweigh the reduced individual freedom and risk of participating in criminal acts, including genital mutilation of their own children. It is an indisputable fact that the need for an individual to belong to a group can be so strong that it even dictates the clothing they wear, either fashion or tradition, as long as accepted within the group. (A certain degree of kinship with the avian world where both feathers and dance can be instrumental at the individual level.) In a seemingly natural manner, also career within any group requires a person to adhere to the opinions established inside the group. In a similarly natural manner, development at the individual level will unfortunately be held back. There are, fortunately, individuals who are prepared to risk their careers for their opinions. One example; The Nobel Prize in chemistry was awarded in 2011 to a man who, in opposition to his boss's orders, proceeded to make a revolutionary discovery about crystals.

Inter-group dynamics and interactions with uncritical individuals within the group

The need for an individual to belong to a group co-exists with the group's need to keep the group homogenous. The latter entails that deviants will be ousted. They, and outside groups may face bullying, murder, ethnic cleansing and wars. The sense of security in a group, a nation, results in rationalization of participation in crimes against humanity as legally and morally acceptable, wiping off all reflections of any possibly upcoming future international criminal tribunals. And we all are familiar how the "group" of scientists wanted to burn Galileo Galilei at the stake, and how Charles Darwin was ridiculed. These obedient pack animals were not using their IK during these events to study facts and to question the pack leader. It is more comfortable to embrace incomprehensive religious conventions and illogical theories.

Interactions between the group and an uncritical profit-hungry mass media

There is also a strong inherited need in these pack animals to flock to the originator of the opinions and ideas, that define the group, and who the mass media has raised to that role by their uncritical worshipping. Mass media's role in this context is not good, as they might also contribute to spreading misleading information. For instance, they write about 'Big bang' as it really has happened even if it is nothing more than a speculative theory. But it sells. What does not sell is what one of the Nobel Prize 2020 winners thinks about the said theory. Furthermore, not publishing the voice of those who doubt in silence is a general problem. As a result falsehoods can be marketed as the truth undisturbed. It has reached the point where school kids are tricked to believe that the above theory has been proven true, which of course is not the case. Why not a word is heard about the "Steady State"?

Wrong becomes right when everyone is doing the wrong

An interesting philosophical question is whether an individual in a group can be held accountable for his or her opinions and actions, or the lack thereof. Our history has shown that the larger the group the smaller the risk, because wrong becomes right when everyone is doing it. Believing in the Lorentz's factor is a good example. A better term to describe this phenomenon is "brain washing".

Group leaders have developed during the years a hidden agenda to stay safe

It is natural for a group's forerunner to fight any dangers threatening their position, because the person with most to lose if a group breaks up is the leader of that group. Consequently, this book will be criticized even if no factual errors could be identified in it. Them trying to keep the group together is more of an egoistic act than an act aiming to secure the group's well-being. Warlords do not give in unless forced to and are, as we know, ready to sacrifice group members to survive. How is it possible to trust leaders who keep plans, hidden behind façades and propaganda, to protect their own careers and safety at the expense of ideology? When there is no longer a need for a group, individuals can be tricked to keep it alive.

Group leaders combat any development threatening their security

The likelihood that particle physicists will read this book is as great as the priesthood at the end of the 1900s reading and getting familiar with Darwin's published theories of evolution. The party understanding the reason in the evolutionary theory was not the clergy. Similarly, present-day authoritarian figures in the field of particle physics will not be the party showing the way forward. Furthermore, it is easy to realize that a strictly regulated and tightly controlled research environment will not result in paradigm shifts, that are necessary for continued development.

The backward group leaders had to eventually give up

The history tells in clear terms that every obscurantist has met their time and dead-end and been forced to capitulate. For those opposing the evolutionary theory, it took just 50 years before they gave up. To make the modern-day particle physicists to seek their way out from the trenches and labyrinths, they are lost in, might take a longer time because most of us others are seemingly indifferent to and guard us against anything based on poor logic. And way too many, who often are equipped with less intellectual capacity, continue to be seduced by these "medicine men". **To hold on to theories that are based on an incorrect interpretation of one experiment performed in 1887 is untenable!**

The layers of fantasies and dust, that have covered both the truth and the logic, were dusted off in 2011 by the first edition of this book. Because of this, and maybe also thanks to all of you, who have read and understood this book, the researchers in the field must, sooner or later, give up their arrogance and capitulate. Mass media and those consuming it will follow, hopefully without the delay so typical to the acceptance process of Darwin's evolutionary theory.

Therefore; the future looks bright for particle physics.

The Great Battle over Common Sense.

– "Nothing moves faster than light!", said the old dragon.
– "Oh yes, something does! Two of your photons meeting each other do travel at double the light speed", said the whistleblowing and truth-seeking Knight.

Lorentz was one of those trying to explain the result from Michelson's and Morley's 1887 experiment. And he has likely stretched the mathematics a bit too far with his solution stipulating that time passes slower for an object moving at a high speed and that the object shrinks. However, he failed to solve the problem despite his so-called Lorentz's factor. Meanwhile the factor captured the interest of a day dreaming dragon. He inserted the factor in various mathematical calculations but kept hitting the dead end. Finally, he decided to introduce two postulates of which one stated that we cannot have one common and absolute reference system in the universe. The second, due to the first: Light quanta (later re-named to photons) are particles with constant speed regardless the speed of the light source. **In other words: Instead of finding a solution to the result obtained from the above experiment, Einstein created an explanation based on fantasies.**

The dragon found this funny, in his unscrupulous ways. He put aside his old and lulling office work that he was never interested in and began to daydream of journeys in time. Maybe he could meet his forefathers, the once-upon-the-time rulers of the world. To illustrate his theories, he inserted his postulates and Lorentz's factor in formulas, and published them; the "Emperor's new clothes".

The gullible mob cheered, but the scientists were baffled how the dragon could so preposterously get away with so brazen and scornful concept of logics based on illogical postulates that are fictional and tailored to solve a riddle. It should be a punishable offence. But the dragon was basking, and the media made money on his daydreams because people have always loved fantastic "truths". However, there was no money in publishing deep and well-founded arguments against the dragon's theories.

The scientists, being older than the dragon, had all passed away after a few decades, and with them the protests passed, too. The theories found their way to schools where pupils got caught in the dragon's web of mathematics woven with pretty, Greek alphabets. Deep behind the mathematics were the two ugly postulates together with the highly questionable Lorentz's factor. The postulates gained a status of an unquestionable truth once the last of the critics had died. Past this point, it would not have been such a good idea for a critic's career to question the postulates. This is because there is a strong human need to establish truths and rules to be abided by. Disagreements in, say in religious matters, lead often to wars. Yet, for two thousand years, people succeeded to agree on Aristoteles's teachings being the only truth. At the time the Knight arrived everyone was kneeling before the dragon. The footmen, keepers of the dragon's teachings in his stronghold of arrogance, were nervously peeking down to the courtyard. Their future was in danger.

– 1 + 1 = 5, said the Knight.
– Not so fast, said the dragon!
– Oh yes, it is! I have just completed a postulate stating that the second occurrence of number 1 equals to 4, said the Knight.

The dragon understood that he had been exposed. The dragon had been smart and opportunistic enough to hide his own growing doubts, always present as a constant shadow. The very thought of doing deskwork or working a conveyor eight to five scared him enough to keep up playing along. But now he began to waver, and his footmen, those who had kept him alive, were terrified.

– I knew you would show up one day, the dragon said.

He was in fact relieved. Despite him being an opportunist with underdeveloped conscience, the last hundred years had been hard on the little conscience he was in possession of. It would be quite nice to retreat to a peaceful place, get to meet his forefathers, but this time without the ephemeral time machine, the dragon pondered. For him to publish his theories had been nothing but challenging the scientists' reactions. Already during that time epoch, the media had begun to steer the mob, and the effect surprised even the dragon. He had initially thought of retreating after the scientists' negative reactions, but instead he let himself be seen in a royal chariot pulled by the media. They made money of him, after all. It was indeed a commercial benefit for them to have the uncritical journalists at their disposal.

The Knight had the dragon, and continued without protests in his monologue:

– And the story that the photon has mass only when in motion just that shows that it is not a particle; it is a wave. The mass consists of the mass in the medium that is the wave. It applies to water, too. A wave in a water has no weight when the water is still.

And a wave has "double" energy because in addition to moving in the direction it is spreading it also move up- and downwards. If we compare the formula for kinetic energy, $E = 0{,}5(M * H^2)$, to the mass-energy equivalence formula, $E = M * C^2$, where C = speed for "your" photons, then the latter formula shows that light is a wave because here we do not have the scale factor of 0.5.

So why not accept the simplest explanation, such as the speed of light is constant, because light moves as waves through a medium. And bringing up your famous thought experiment with the train and the light – you should see the light as a wave moving through the dark matter that is bound to the surrounding matter. In this way, the concept becomes logic, the Knight said. (The thought experiment is described in the book.)

The Knight had already solved the riddle from the 1887 experiment when he had understood that the ether, the dark matter, was bound to the adjacent matter, and that it did not allow to be affected by the gravitation, something he also later demonstrated. The discovery contributed to the conclusion that the light consists of waves traveling through the ether. This aligned also with the results obtained from experiments conducted by universities and institutions. And because he was able to demonstrate that the liquid substance called Helium II was weightless, further supporting his conclusion on the ether's relation to the matter, the Knight became so enthusiastic that he published his discoveries in a book. But instead of being cheered by people he was met by a wall of silence. It became clear that no one, except for a few, was interested in reading the book during the next ten years even if the subject often came up in mass media and therefore it should have been of public interest. Once he gave the book to one of the dragon's footmen, whom he met during a train journey, to read. The Knight promised him a sizeable payment if he could find one error in the book. The reason why the footman never got back to him remained unclear. And because none of the other footmen in the four corners of the country never responded to the Knight's letters, he called a female professor emeritus and asked whether she might be willing to review the book. Her reaction was by far the most unfriendly so far. The dragon's footmen are short of a few important characteristics needed to become something else, and as such they are perfect in their roles. The Knight had no help from the public either. Despite not understanding the dragon's theories, they rather believed in his fantasies than in the Knight's logic. The paradox can only be explained by effective brain washing by the media that has always made a living by selling papers with tributes to the dragon.

The Knight could not count on getting help from anyone, and had no other option than to meet the dragon alone in an attempt to get him to retreat. He had hoped to avoid this as he had begun to develop a sense of understanding the dragon who, in opposite to his footmen, had at least imagination. His intelligence may not have been at the level of his imagination but together with his charm, it had been sufficient to trick the public to swallow any postulate. And the ultimate power behind his success in spreading his dreams about a more interesting world was the media. It was an unfortunate but pardonable situation because the media is constructed to make money, and lies sell. But now, there are more money to make from the story of a well-established falsehood, a scandal, an unforgivable and permanent disgrace, that is, this story that has been waiting for a century.

Buzzing from the dragon's footmen woke the Knight from his thoughts. He stretched and continued:

– And the cosmological constant that you inserted to prevent the formula from describing a growing universe? You removed it upon learning that the universe is expanding. But astronomers re-inserted it to represent the dark energy that you failed to include in your calculations! And the dark energy should be the strings in motion in the dark matter, shouldn't they. Had you just accepted that light consists of waves moving in the ether, then you would have understood it.

And the error in your mathematical calculations that made black holes endlessly small, it must have been caused by Lorentz's factor? How can you trust any theories that contain this factor?

– But not everything can be incorrect, the poor dragon said, playing his last card. For example, photons. It is obvious that photons are particles being affected by the gravitation because light is bent by stars.

– Nonsense! Light that passes near a star goes through dark matter that is linked to matter, and a wave changes always direction when the medium it is propagating in changes, if its angle is not perpendicular, said the Knight. And the reason why black holes are black is not due to gravitation keeping the photons from leaving – black holes are black because no nuclear or chemical reactions are possible in crushed atoms.

And if you still believe that your photons are particles then you must be dump enough to believe that one photon knows what another photon is going to do. An experiment, where photons are assumed to

be particles, have indeed reached that conclusion. By assuming instead that the photons are waves moving through the dark matter, the result from that experiment become logic.
The dragon felt ashamed. The Knight's comments had hit hard. And now the last words uttered by the Knight penetrated like a death blow through the dragon. He got confused and in his daze, he swallowed nervously, and burned up. The fight was over. The dragon had been killed by its own fire, dark energy. There was confusion and chaos. The dragon was dead, and the footmen took off fleeing. You could hear the silence. It was the same silence you hear when a train stops and waits before it begins to reverse to hit right tracks. The exhausted Knight found his way back to the castle to free the immortal Princess Truth. The only entertainment she had been able to enjoy through all these years behind the locked door was to watch the footmen using both crutches and bandages in their growing efforts in keeping the stumbling falsehood alive. From her life-long experience, she knew that one day she would be free again. This time, though, she was not looking forward for all the work needed to correct the massive disarray left behind by the footmen.

In fact, the Knight's input was not necessary at the end, because the dragon's own men had already realized what the Knight wanted to demonstrate. Unfortunately, their fear of the dragon was too deep for them to dare see that they had already confirmed the Knight's theories with a rapidly alternating magnetic field. They had created recordable waves moving through the dark matter, waves that they named 'virtual photons' patching up a poor theory about the so-called photons with another patch.

The Knight came to think of that he should have applauded the dragon for his research on the photoelectric effect, even if was assisted and close to not being the first to come out with the results from this research. It would have been unnecessary to hurt the dragon's feelings by revealing that he disagreed on the conclusions. The Knight's explanation to this effect is based on his discovery of:
1) the string chains in the dark matter keep protons and electrons together, and 2) the light quanta, or the photon, as it is now called, is a wave moving through the dark matter. The results from 1) and 2) are as follows; based on the atom's configuration, i.e., how an electron is attached with one or more string chains, equivalently different energy levels of the light (wave motion) are required to detach an electron from its atom.

Once the Truth was freed, the Knight had hung up his armor to retire. There was nothing more he could do. The Truth had to continue the battle alone. And as she had feared, this was a battle against the boneheaded silliness that would take years. Deep pockets of resistance remained in many of the strongholds of arrogance where footmen were fighting for their life. Their future was far from bright. The high priests besieged under dark and stormy skies were the most exposed. The sun that always shines after rain and storm may not do it there, the Knight thought.

Following the dragon's death, he had with lightness in his steps wandered back to his house to write a new edition of his book about chasing the zero gravity. Maybe Truth could have use of it although this seemed uncertain since people only read simpler topics that interest them personally. Other topics are absorbed unquestioningly as hearsay, prejudice or through headlines, and the majority will happily steer away from any efforts to gain knowledge. Thus, to wake the interest in readers, one must appeal to their primitive needs. By succeeding to get Princess Truth to undress to expose the naked Truth to people, then I think I would win the battle, the Knight thought. Suddenly, he was taken over by panic. If people aren't aware that the dragon is gone and if they are dumber than what he is wise, then the battle is lost. The years-long voluntary mission had been his calling. And now the only thing he could do was to wait and see whether passing of time would cease the unwarranted worshipping of the dragon.

The Naked Truth.

Those who read and understand the content of this book will see the definition of the four natural forces (gravity, magnetism, and the two types of nuclear forces inside an atomic nucleus) as different sides of one and a same thing. This is because separating the four natural forces will not get to the bottom of the truth. Because everything behind these forces is about strings of the dark matter that, in their constant motion, either latch to each other or bounces off each other. In other words, **movements of the strings, the energy of dark matter, is the only natural force** that is behind everything. This is the simple, completely undressed and totally naked truth.

The black matter's never-ending noise, caused by the strings in motion in and around us, can be compared to the chaotic bouncing of atoms and molecules in something like air. Movements can be propagated from one string to another string, much like a wave. When strings collide, by random event they either get latched to each other (attraction) or they bounce off each other (repulsion). If they latch to each they form string chains and both "dark" and subatomic particles. And if the strings bounce off each other, this will result in new waves which lead to new collisions, etc. Since there are different types of particles one can assume that there are different types of strings.

Through their turbulent lives, strings are interacting also with the subatomic particles. An electron can indeed latch to a string chain, or the power of a wave moving through the string mass can force off an electron from one or more of the string chains that couple it to an atom. A wave can also just bounce against an electron and the string chain coupled to the electron, which occurs when the wave is too weak to knock loose the electron. The described interactions will create new waves, light with wave lengths that identify the atom in question. The number of so called light quanta needed to knock loose an electron is proportional to the number of string chains securing the electron in place. Therefore, the energy needed for this are different levels. And the energy from string chains connected to an electron will manifest itself in specific areas in the color spectrum where shorter wave length amounts to higher energy. (By this type of interactions, atoms are built up.)

As described in the book, an electron is secured to an atom with one to four string chains. One of these string chains links the electron to a proton, and the other three string chains link the electron to the adjacent electrons. In inert gases with eight electrons on their outermost shell, the atom is cube-shaped with one electron in each corner and string chains forming the sides. Consequently, these electrons are linked where they sit, with all three string chains between each other in the same atom. This is why atoms in inert gases are unable to form links to other atoms. At the same token this explains why other atoms merge to molecules to achieve eight electrons in one of the atom's outer shells. Chemistry is thus being controlled by particle physics. Have the electrons had a different number of handles reaching towards string chains, our world would have been made of other molecules, and the life as we know it would not exist.

The puzzle consisting of logical pieces of particle physics, is completed. The book illustrates how well all the puzzle pieces fit, where found

at the bottom of the *"The Naked Scene of Creation"*.

The discovery and defense of the naked scene of creation.